# THE NEXT GREAT MIGRATION

# THE NEXT GREAT MIGRATION

The Beauty and Terror *of*
Life on the Move

## SONIA SHAH

BLOOMSBURY PUBLISHING

NEW YORK · LONDON · OXFORD · NEW DELHI · SYDNEY

BLOOMSBURY PUBLISHING
Bloomsbury Publishing Inc.
1385 Broadway, New York, NY 10018, USA

BLOOMSBURY, BLOOMSBURY PUBLISHING, and the Diana logo are
trademarks of Bloomsbury Publishing Plc

First published in the United States 2020

ISBN: HB: 978-1-63557-197-4; eBook: 978-1-63557-199-8

LIBRARY OF CONGRESS CATALOGING-IN-PUBLICATION DATA

Names: Shah, Sonia, author.
Title: The next great migration : the beauty and
terror of life on the move / Sonia Shah.
Description: New York : Bloomsbury, 2020. |
Includes bibliographical references and index.
Identifiers: LCCN 2019042062 | ISBN 9781635571974 (hardcover) |
ISBN 9781635571998 (ebook)
Subjects: LCSH: Emigration and immigration—History. | Emigration
and Immigration—Government policy. | Immigrants—Social conditions. |
Refugees—Social conditions. | Global environmental change—Social aspects.
Classification: LCC JV6201 .S49 2020 | DDC 304.809—dc23
LC record available at https://lccn.loc.gov/2019042062

2 4 6 8 10 9 7 5 3 1

Typeset by Westchester Publishing Services
Printed and bound in the U.S.A. by Berryville Graphics Inc., Berryville, Virginia

To find out more about our authors and books, visit www.bloomsbury.com
and sign up for our newsletters.

Bloomsbury books may be purchased for business or promotional use. For
information on bulk purchases, please contact Macmillan Corporate and
Premium Sales Department at specialmarkets@macmillan.com.

*To the transplants and the dislocated and those still en route,*
*who poured me tea and shared their stories*

No other mammal moves around like we do. We jump borders.
We push into new territory even when we have resources
where we are . . . There's a kind of madness to it. Sailing out
into the ocean, you have no idea what's on the other side.
And now we go to Mars. We never stop. Why?

—Svante Pääbo


They did what human beings looking for freedom,
throughout history, have often done. They left.

—Isabel Wilkerson


Borders? I have never seen one. But I have heard they
exist in the minds of some people.

—Thor Heyerdahl

# CONTENTS

# LIST OF MAPS

I

# EXODUS

The early spring sky is a deep saturated blue next to the muddy brown of the dry, scrubby hills of the San Miguel Mountains in southern California. Save for the thrum of a distant bulldozer, this open, unassuming place is quiet. It's visually calm, too, with few dramatic features: just sandy, sunbaked ground, gentle slopes, and low shrubs and grasses of varying russet shades. The hills seem to continue indefinitely into the distance, crisscrossed by rutted dirt roads and thin walking trails.

The creatures I've come to see are equally unassuming. *Euphydryas editha*, aka Edith's checkerspot butterflies, are so slight and unobtrusive that they would be barely detectable in any amateurish photograph I might shoot on the iPhone tucked into my back pocket. The plant they live and feed on, the dwarf plantago, is equally modest. It grows just a few inches high, with needle-like leaves and tiny, translucent white flowers on thin stalks. It's about as dazzling as dry grass. You could easily crush it underfoot—as I did—without noticing its presence at all.

The butterfly expert accompanying me—the perfectly named Spring Strahm—brought me here, bouncing along in her four-wheel-drive truck, over roads that had been closed to the public since 2015. Finding a checkerspot butterfly in these mountains, she tells me, is "kind of like seeing a unicorn," but she's famously good at it.

We amble slowly into the hills, Strahm occasionally dropping to hands and knees to inspect some low-growing grasses for hidden butterflies and turn over a few leaves in search of caterpillars. After the better part of an hour, having acquired nothing beyond a few rivulets of sweat, she decides we've had enough. It's time to head back to the truck and try to find the elusive checkerspot elsewhere. I unscrew my water bottle for a quick swig, adjust my backpack, and follow her back down the trail.

A few minutes later, she stops short. She stands there unmoving, blocking the path. Then I notice she is staring at the wizened hiking boots on her feet. I look down. A low fluttering cloud of butterflies hovers around our ankles.

~

I'd come to see the checkerspots thanks to Camille Parmesan. With her mane of dark curls and ice-blue eyes, Parmesan could pass as an earthy, compact version of Wonder Woman, if Wonder Woman liked dirt and bugs instead of lassoes and invisible jets and spoke in regional slang. Parmesan grew up in an Italian family in Texas. She freely uses the word *ain't* and prefers *honker* over *large*, and *out the wazoo* over *abundant*.

Parmesan first started studying checkerspots as a graduate student in ecology in the 1980s, after giving up on the study of birds (they wake up too early), lab-reared primates (too unnatural), and honeybees (too many stings). She liked butterflies, she says, because they were easy to watch in their natural environment and amenable to manipulation. She'd spent her childhood camping out with her mom, studying field guides and identifying plants and birds. Her mother loved botany but by profession was a geologist, working, like many in Parmesan's Texan family, in the oil industry. She'd provided her daughter with a uniquely geological spin on her camp-side botanical teachings. From her, Parmesan had learned about the deep history of wild species through geological time, about how they'd advance northward during warm periods and then retreat during cold ones, rising and falling along with the ice ages.

By the time she entered the world of checkerspot biology, conditions were dire for the field's diminutive subject. She knew, from dusty museum records and the prodigious personal collections of amateur butterfly enthusiasts, that the checkerspot had once been common, with colonies up and down the mountainous west coast of North America, from Baja California in Mexico to British Columbia in Canada. Legend had it that one enterprising butterfly collector had caught masses of them just by riding his motorcycle along the coast, one arm extended, butterfly net in hand. But for years their numbers had been declining.

The reason was pretty clear to most ecologists. The checkerspot was not able to really move much. As fuzzy black caterpillars, they rarely inched more than a handful of feet from the plants from which they hatched. Even after they unfurled their spotted wings, they stayed low to the ground and close to home, rarely flying more than a few meters from the scenes of their metamorphoses. Wind or rain would send them clambering with their thin spindly legs to the base of their dwarf plantagos, settling as low to the ground as possible to prevent their delicate bodies from being inadvertently swept away in a gust. They were widely known, in the field, as "sedentary," the entomological equivalent of homebodies.

Meanwhile they were getting squeezed. The dwarf plantagos they preferred were drying out in the southern part of their range, as the carbon-torched climate in northern Mexico grew hotter and drier. The urban sprawl of growing cities such as Los Angeles and

3

San Francisco, meanwhile, swallowed up the gentle, sun-drenched slopes of the northern end of their range. Trapped between climate change on one end and urban expansion on the other, the checkerspot, most butterfly experts believed, was doomed.

It was a pretty simple story, being told in a range of variations across the globe. Parmesan had no illusions about changing the basic plotline, but she thought she might be able to document the specific ways in which the butterfly responded to the pressures it faced. A few of the colonies might exhibit some subtle local adaptation, perhaps, or emit some striking signal before their inevitable collapse. If she conducted a proper census, then crunched the data, with some sophisticated statistical analyses she might be able to scrape a passable dissertation out of it. Her research would be, in a way, an elaborate documentation of a species' death throes, but that's what a lot of ecology had become in this age of mass extinctions. There were worse ways to get a PhD.

Plus, the butterflies hatched in glorious spring weather, didn't wake up until ten A.M., and were most easily spotted on sunny windless days. For four years, Parmesan spent her summers driving up and down the West Coast, hunting for butterflies by day and camping out in the mountains by night.

She didn't have particularly high hopes for her results—"I wasn't sure I would come out with anything at the end of it," she says. Then she started analyzing the data. The butterfly's numbers had contracted, compared to the historical records, which was what she'd expected. But there was something more, too: a signal in the noise, one that would upend her career and draw the attention of journalists like me from all over the world.

"I start looking at the pattern," she told me when we met at a Tex-Mex restaurant in Austin. "And I see that the extinction rate is really high in the south, and really low in the north and in the mountains. I was expecting this complex pattern, and I thought, this is really simple! . . . I couldn't have gotten clearer data."

Like the wild species her mom had told her about on summer camping trips years ago, the butterfly had responded to the changing climate the way wild species had in millennia past.

It had moved.

"It's just shifting its range northward and upward!" she says. The finding, now more than two decades old, still fills her with surprised delight. She gathers her hair with both hands and tosses it behind her back with a little shimmy. "My goodness!"

~

Parmesan published results from her butterfly survey in 1996. At that time, only two other studies had documented a wild species shifting its range in response to climate change, one in plant communities on the tops of mountains in the Alps and another in sea stars and mussels in Monterey Bay. Those were "very good papers," she says, "but very small areas." They could be easily dismissed as anomalous. While life-saving movements in response to climate change seemed theoretically possible at the time, few scientists dared to hope that wild species would be able to accomplish them at any meaningful scale.

Parmesan's study of checkerspots, in contrast, showed a consistent pattern of movement across half of North America. She got a coveted single-author paper in the prestigious journal *Nature* and instantly ascended to the top ranks of climate change science. She became a member of the United Nations Intergovernmental Panel on Climate Change, a position that allowed her to review nearly a thousand other ecological studies, searching for the same signal she'd found in the checkerspots. Indeed, the butterfly's poleward shift was no anomaly. The same pattern could be found in fifty-seven species of butterflies in Europe. And in marine organisms. And in birds.

Scientists who studied everything from plankton to frogs started reexamining their data. They found that of the four thousand species that they'd tracked, between 40 and 70 percent had altered their distribution over the past handful of decades, around 90 percent into cooler lands and waters in sync with the changing climate. On average, terrestrial species were moving nearly twenty kilometers every decade, in a steady march toward the poles. Marine creatures were moving into cooler waters even faster, moving about seventy-five kilometers per decade on average. Those averages obscured some spectacular leaps among specific creatures. Atlantic cod, for example, had shifted more than two hundred kilometers per decade. In the Andes, frogs and fungi species had climbed four hundred meters upward over the past seventy years.

Even the most seemingly immobile wild species were on the move. Coral polyps, which over decades form the branching thickets and sprawling nubby plates of the world's coral reefs, may seem the picture of stately immobility. They are literally stone walls, absorbing the fury of the open ocean, protecting millions of fish species and seaside communities. And yet the coral reefs are moving, too. Scientists peering through glass-bottomed boats had been surveying corals around the islands of Japan since the 1930s. In 2011 scientists discovered that two species in particular—*Acropora hyacinthus* and *Acropora muricata*—had been moving northward at a speed of fourteen kilometers every year.

In the meteorologist Edward Lorenz's famous formulation, the flapping of a butterfly's wings creates a minor atmospheric disturbance that, because of the complex interplay of interconnected factors, ends up altering the path of a distant tornado. That was a poetic metaphor for one of my favorite insights, that small changes can have unexpectedly large effects. The whole point of the metaphor is that a butterfly's flight is a seemingly insignificant factor, but still I figure he must have had something like the majestic intercontinentally migrating monarch butterfly in mind when he coined that turn of phrase. He couldn't have been thinking of checkerspots. Having met a few of these butterflies and witnessed their unimpressively slow, low flying, I doubt their collective flapping could cause even a whisper of breeze, let alone any kind of major meteorological event.

And yet the little butterfly had triggered an outsized effect of a kind, its unlikely journey lifting the veil on a dramatic global phenomenon. In Unalakeet, on the northwest coast of Alaska, hunters find parasites from more than 950 miles southeast in British Colombia squirming under the skin of the wild birds they hunt. Red foxes spread north into Arctic fox territory. In Cape Cod boat owners encounter manatees from Florida casually sipping water from drainage pipes at their marinas.

A wild exodus has begun. It is happening on every continent and in every ocean.

~

The towering Dhauladhar mountain range, with its eighteen-thousand-foot peaks, looms over the precariously perched village of McLeodganj, nestled on a forested ridge nearly seven thousand feet up in the foothills of the Himalayas. I arrived after a terrifying twelve-hour taxi ride from New Delhi. My driver, used to the sea-level pressure of the plains and dressed in a crinkled short-sleeve cotton shirt, was dizzied, cold, and fed up by the time we reached the center of McLeodganj late that night. He pulled over and discharged us, along with six months' worth of luggage, in the middle of the village square, several vertiginous kilometers from our hotel, and fled.

It seemed unforgivable at the time, but my attitude softened the following morning, as the mist burned off, revealing the town's heart-stopping panoramas. The Himalayan pine trees that cling to the mountainsides abruptly peter out in the rocky upper reaches of the peaks, creating a natural border known as the "tree line." Above the line rise barefaced cliffs, streaked by narrow waterfalls. Just hauling my body around at this altitude required substantive feats of navigational prowess and physical stamina. I wouldn't have wanted to try driving a rickety Delhi taxicab here either. The narrow, unmarked alleys were steep, the air thin, and deathly, unfenced precipices appeared around every corner. I arrived equipped with the latest in mountain gear, purchased at great expense from specialty shops for my brief stay in these mountains: a nylon jacket coated in polyurethane, sturdy waterproof hiking boots, special sweat-wicking woolen socks. They didn't do much to reduce the overwhelming feeling of being unprepared for the forbidding landscape. I huffed and puffed along the trails above the town, thankful that the sole witnesses to my growing discomfort were the rhesus monkeys scampering through the pines above, and the friendly local dogs that followed patiently behind me.

If any geographic feature should arrest movement, it is the Himalayas. They form an impassable wall, geographically speaking. On one side, the frigid air of the north collects, barred from reaching the tropical southern plains below. On the other, approaching monsoon clouds smash into the peaks, dropping their liquid interiors on the ridges as if released by a recently opened sluice.

And yet even here, up against this giant wall, living things inch, drift, and climb, untethered to any permanent anchors. Every year the young saplings in the forests establish themselves a little bit higher up the slopes. When curious scientists marked a transect and measured the age of the trees along it, they discovered what was happening. Since 1880 the forests had steadily climbed the mountainside, moving nineteen meters uphill every decade. They bring with them the rhododendrons and apple trees, and the insects that live in and on them. People in Tibet, a high-elevation tundra on the northern side of the Himalayas, first reported suffering from strange itchy bites in 2009. It was the first time anyone there could ever remember being bit by a mosquito.

People are on the move here, too, their migrant tracks wending into the valleys, around the curves of the mountainsides, and over the high alpine passes of the Himalayas. More than a hundred thousand people from the Tibetan plateau steadily trickle in to McLeodganj some five hundred miles away, fleeing the Chinese government's persecution and repression. Many are Buddhist monks and nuns who followed the fourteenth Dalai Lama, who arrived in 1959 and now live in a run-down temple complex just down the narrow winding road from my modest hotel. I saw them in their bright saffron robes sipping cappuccinos in the local cafés and amiably ascending the steep rocky trails around town in their simple sandals and woolen shawls, chatting to one another on smartphones tucked into their robes. Unlike me, who arrived via airplane and taxicab, they had walked over the mountains to get here. Each of their perilous journeys over glaciers and high mountain passes took a month.

~

The news today—on any day—is full of stories of people on the move. African migrants fleeing starvation and persecution cram themselves onto leaky boats to cross the Mediterranean. Afghans and Syrians wilting in tattered camps are herded back to the bombs and beheadings they've fled. Women hauling toddlers on their hips walk hundreds of miles from Honduras and Guatemala to reach the U.S. border. As I write this, my phone buzzes beside me with breaking news: the governor of Florida has ordered the evacuation of more than a million Floridians, as a category four hurricane approaches, threatening disaster. The roads on the peninsula will soon be swarming with families seeking higher ground.

The movements of wild species are shaped primarily by the constraints of their own biological capacities and the particular qualities of the geographic features they encounter on their journeys, such as the steepness of mountainsides and the speed and saltiness of ocean currents. The paths taken by human migrants, in contrast, are shaped primarily by abstractions. Distant political leaders lay down rules based on political and economic concerns, allowing some in and keeping others out. They draw and redraw invisible lines on the landscape in biologically arbitrary ways. Transportation companies offer passage on certain routes and not others, depending less on the wind, weather, and tides than on which ones net them the highest margins of profit.

We move, nevertheless. More people live outside their countries of birth today than at any time before. The reasons vary. Between 2008 and 2014, floods, storms, earthquakes, and the like sent 26 million people into motion each year. Violence and persecution in unstable societies stir other journeys. In 2015 over 15 million people were forced to flee their countries, more than at any time since the Second World War. For every person who crossed an international border, there were more than twenty-five others whose peregrinations had yet to impinge on one of those invisible lines. All these specific flows collapse into a broader one, shifting our populations from the countryside into the world's cities. By 2030 the accelerating movement of people into metropolises will result in the majority of us being city dwellers for the first time ever. And the extent of our movements is likely to grow for years to come. By 2045 the spread of deserts in sub-Saharan Africa is expected to compel 60 million inhabitants to pick up and leave. By 2100 rising sea levels could add another 180 million to their ranks.

These statistics, eye-popping as they are, offer only a partial snapshot of the scale and pace of our current era of migration. There is no central authority that collects data on human migration. People who cross international borders may get recorded by some authorities on one side or the other, but only in some places and some of the time.

Authorities mostly count who's coming in, shielding their eyes from the parades of people who leave. Many people on the move try to escape official notice, traveling furtively undercover; or they move within borders, avoiding surveillance altogether. Government officials may try to estimate the number of people who cross their borders without permission, but the best they have are estimates, based on fragmentary evidence: the number of people border authorities catch in the act; the number of people caught in the act who admit they will try to do it again; the number who do, in fact, try again and are caught once more. Whole categories of human migrants—those who go back and forth over borders, for example, for seasonal work or harvests—are not included in any official statistics.

Given all this, the true number of human migrants is not fully knowable. But the central fact is clear: like our wild cousins, people are on the move, too.

Over the past handful of years, as the climate's grip on how we move has become increasingly apparent, now evidence of the centrality of migration in our biology and history has emerged. New genetic techniques have revealed how deep into the past our story of migration runs. New navigational technologies have uncovered the scale and complexity of both human and wild movements around the planet. While our coming migrations may not proceed fast enough to keep pace with our shifting climate, a growing body of evidence suggests they may be our best shot at preserving biodiversity and resilient human societies.

~

The next great migration is upon us. The trouble is, from the earliest years of childhood, we are taught that plants, animals, and people belong in certain places. It's why we call the goose the "Canada" goose, the maple the "Japanese" maple. It's why we use the camel to represent the Middle East and the kangaroo to stand for Australia. It's why we use our imagined or known continental origins as shorthand to describe ourselves in everything from our social interactions to our medical forms: we are "Americans," or "Africans," or "Asians," or "Europeans," a centuries-old marker encoded visually in the color of our skin and the texture of our hair, regardless of where we might happen to live.

By describing peoples and species as "from" certain places, we invoke a specific idea about the past. It traces back to the eighteenth century, when European naturalists first started cataloging the natural world. Assuming that peoples and wild creatures had stayed mostly fixed in their places throughout history, they named creatures and peoples based on those places, conflating one with the other as if they'd been joined since time immemorial.

Those centuries-old taxonomies formed the foundation for modern ideas about our biological history. Today a range of fields from ecology to genetics and biogeography allude to long periods of isolation in our distant past, when species and peoples remained ensconced in their habitats, each evolving in their separate locales.

This stillness at the center of our ideas about the past necessarily casts migrants and migrations as anomalous and disruptive. Early twentieth-century naturalists dismissed migration as an ecologically useless and even dangerous behavior, warning of "disastrous results" should migrant animals be allowed to move freely. Conservationists and other scientists warned that human migration, too, would precipitate biological calamity. The most predictable outcome of human migration—sexual reproduction between people who traced their ancestry to different places— would result in degenerated, mutant hybrids, leading scientists proclaimed.

The free movement of peoples would allow hungry hordes of foreigners to overrun the country, postwar population biologists said, pointing to their studies of population dynamics in butterflies and rats. Would-be human migrants, one wrote, would not "starve gracefully." They'd migrate, to our ruin. Wild species on the move, late twentieth-century ecologists added, would trigger "environmental apocalypse."

These ideas about migrants and migration were often based on flimsy evidence: mysterious female body parts that don't, in fact, exist; hybrid monsters that have never been found; a storied spectacle of wild migrants leaping into the Arctic sea that never, in fact, happened; a phenomenon of crazed aggression and voraciousness produced by crowding that doesn't actually transpire. For decades, they suppressed the truth about the promise of migration, regardless. Geneticists who discovered the fact of our common migratory history minimized its extent. Biogeographers puzzling over the wide distribution of species and peoples across the planet dismissed the possibility of active movement, presuming instead that ancient geological forces passively carried them around.

Scientific ideas that cast migration as a form of disorder were not obscure theoretical concerns confined to esoteric academic journals. They were widely disseminated in popular culture. They influenced the closing of the U.S. borders in the early twentieth century, inspired the fascist dreams of Nazis, and provided the theoretical ballast for today's generation of anti-immigration lobbyists and policy makers.

They roil fear and panic about the next great migration today, reshaping the politics of the most powerful nations on earth. Conservationists warn of the "invasive" appetites of alien species moving into habitats already populated by native ones. Biomedical experts warn of migrant species carrying foreign microbes into new places, sparking epidemics that will threaten the public health. Foreign policy experts predict instability and violence as the necessary result of mass migrations forced by climate change. Antimigrant politicians speak of economic calamity and worse.

~

The idea of migration as a disruptive force has fueled my own work as a journalist. For years I reported and wrote about the damage caused by biota on the move. I investigated how mosquitoes flitting across landscapes and nations infected societies with malaria parasites, shaping the rise and fall of empires, and how cholera bacteria traveling across continents in the bodies of traders and travelers triggered pandemics that reshaped the global economy. The disruptive impact of these microbes out of place conformed to my sense of movement as aberrant, something anomalous that needed to be examined and explained. It echoed that other strange fact that required explanation, the incongruity of my own body in space, unmoored by the movement in my family's past.

My migratory past traces back to the late nineteenth century, to two fishing villages in Gujarat, along the western shores of India. These villages, with their coasts jutting into the Arabian Sea, had first been settled by migrants from Europe, Southeast Asia, and Africa. Since then they'd been repeatedly buffeted by waves of traders, invaders, and colonists who joined the locals: Persians, Macedonians, Mughals, and British, among others.

My great-grandfathers grew up in these villages. One was a hunchbacked peddler who hawked cotton saris; the other owned a small shop selling metal cooking vessels. They both grew up with customs designed to resist the migratory tides around them. One stricture, for example, held that they could marry only into families that followed the same sect of Jainism theirs did and who lived no farther than one village over. Considering that less than 1 percent of the population of the entire state of Gujarat today are Jains of any sect at all, those rules likely made for some pretty slim pickings.

Their sons, my grandfathers, adhered to family custom and married the young daughters of wealthy village families, but that did not stop them from joining the nineteenth century's global migration from the countryside into the newly industrializing cities. One settled in teeming Mumbai, cramming his five children into a two-room flat in a *chawl* tenement, a new kind of building constructed specifically for the working-class migrants like him who had flooded the city. The other went south to Tamil-speaking Coimbatore, where he moved into a small house owned by the company that employed him. There, on a pile of mattresses in a sparely furnished stone-floored room, my grandmother gave birth to eight children, of whom six survived to adulthood. In Coimbatore and Mumbai, these two

# Navigating the Darién Gap

More people live outside the countries of their birth today than at any time in the past. Because of the difficulty of reaching the United States and Canada directly, many migrants travel to South America first and then journey to the U.S. border over land. That requires crossing the roadless jungles and mountains of the Darién Gap, depicted in this map, on foot.

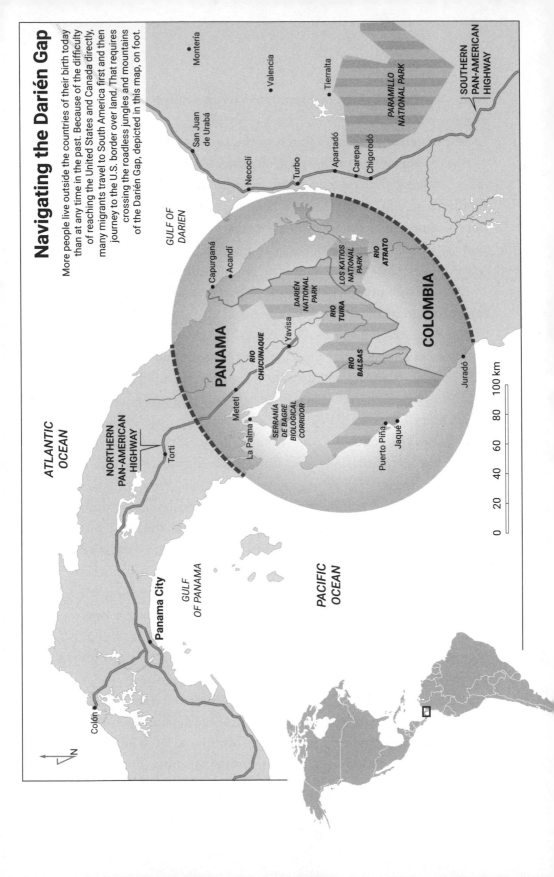

N

ATLANTIC OCEAN

Colón

Panama City

GULF OF PANAMA

PACIFIC OCEAN

NORTHERN PAN-AMERICAN HIGHWAY

Torti

PANAMA

La Palma

Meteti

RIO CHUCUNAQUE

Yavisa

SERRANÍA DE BAGRE BIOLOGICAL CORRIDOR

Puerto Piña

Jaqué

RIO BALSAS

RIO TUIRA

DARIÉN NATIONAL PARK

LOS KATIOS NATIONAL PARK

RIO ATRATO

COLOMBIA

Juradó

Acandí

Capurganá

GULF OF DARIEN

San Juan de Urabá

Montería

Valencia

Tierralta

Necoclí

Turbo

Apartadó

Carepa

Chigorodó

PARAMILLO NATIONAL PARK

SOUTHERN PAN-AMERICAN HIGHWAY

0    20    40    60    80    100 km

now-far-flung families poured their resources into two of their eleven progeny: my mother and father, who each got an education and went to medical school.

A new migratory path opened up just as they graduated. Since the early twentieth century, U.S. borders had been closed to people from Asia, Africa, and southern and eastern Europe, having been deemed by the then-cutting-edge science of eugenics to be mentally defective and biologically undesirable. But the need for medical workers to staff the newly established government programs of Medicare and Medicaid had created an acute shortage of physicians in the United States. Seated at the foot of the Statue of Liberty one crisp October day in 1965, President Lyndon Johnson signed a bill reversing the eugenics-based bans of the past, opening the borders to skilled workers from overseas. A year later my parents had so many offers for medical jobs in New York City that they resorted to evaluating them based on whether they included apartments, and whether those apartments had balconies.

My dad left for the United States first. Six weeks later my mother arrived at JFK Airport wearing a sari and *chappals*, her thin socks bunching up around their single-toe enclosures. They were among four thousand Indian migrants to the United States that year, the vanguards of a new migrant wave.

Today, more than fifty years later, my parents' migration remains the central fact of their lives. It's why they will always long for the perfect mango, why the voice-recognition app on my father's phone will never understand his grammatically perfect English, why they have missed out on countless birthdays and arguments and family dramas. It's why they have been ever since, in some ways, severed from their past, related by blood to people who could no longer make sense of their lives. My grandmother used to cry when she heard that, in America, her son washed the dishes after dinner. In the flat she'd raised him in, dishwashing was a job for the day laborers, who crouched on their haunches on the slimy tiled floors of the common washing area and slept on thin rough mats on the terrace.

I was born in New York City, a few years after my parents' migration, one of more than 4 million descendants of the migratory wave that carried them to the United States. The consequences of that deceptively simple prior event lodged deep in my bones, sending out pangs and throbs like a slightly off-kilter metal implant. On one hand, I was glad to be free of my parents' past. Their transoceanic move had snipped the threads that connected me and my sister to a way of life that I didn't always admire, lifting us away like balloons. I didn't want to have to memorize poems or prostrate myself to elders, as my cousins did, or be the subject of resigned sighs when some future husband, arranged through family connections, agreed with the then-consensus among our Indian relations that some wives needed beating. That was clear to me from a very young age. I remember, as a child, wandering through the high-rise apartment my parents had bought in Mumbai, which they planned to move into after having spent a few years in the United States. With its stunning vista over the sea, the flat vastly outshone our cramped subterranean apartment in Canarsie, Brooklyn. Still, when they decided not to move after all, I felt as if I'd dodged a death sentence.

At the same time, their migration instilled in me an acute feeling of being somehow out of place, one that's taken nearly five decades to quell. As a child, I was ashamed of even small things, like my preference for suspiciously fruity strawberry ice cream over the unimpeachably American chocolate for which the other children clamored. During visits to India, I felt equally ashamed for not tolerating spicy foods and overripe mangoes. Everyone seemed to know instantly I was not one of their own and seemed more than happy to say so. At home, people around me would look at my black hair and brown skin and refuse to accept my residence in various American cities and suburbs as authentic, asking to know where I was "really" from.

For years, I accepted their presumption of my occupation of space on the North American continent as in some way abnormal. Adopting their sense of my oddity, I pushed myself from the center to the margins. I never presented myself as a regular American person, but always some marginal permutation of one: a South Asian American, say, or an Indian American, perhaps. Even after living in Boston for more than a decade, I didn't publicly cheer when the Red Sox won or wail over the city's various tragedies. That felt presumptuous, because I didn't consider myself as being "from" that place, even though I'd borne both my children there. I still don't say I'm "from" Baltimore, though I've lived on the outskirts of this city for over a decade.

I became a migrant myself, for a few years. When my kids were small, my husband and I moved to northeastern Australia, where he had accepted a research job at a university. He hoped we'd stay and even secured citizenship for all of us. But as my sons acquired Australian accents and were subjected to the locals' skewed ideas about race, my enthusiasm for the transcontinental shift—never terribly great—started to flag. I began to understand why my parents had always seemed to lack a certain confidence in their American-raised progeny, as if we were the product of some experiment they'd conducted and they were still analyzing the results. I didn't want to create another rift between the generations. Plus, my father cried on the phone when I called.

After a few years, we left, my misgivings about the turmoil caused by migration intact. It was easy enough to agree with the conventional wisdom, which located the source of that tumult in the migratory act itself and the seemingly contrarian impulses that drove it.

But then I started tracking migrant routes around the world.

~

With his chiseled features, dark stubble, and short, silver-specked hair, Ghulam Haqyar could easily pass as a Hollywood actor. Haqyar worked as a manager for an international NGO in Herat province in the northwestern corner of Afghanistan, enjoying a healthy salary and a comfortable home in Herat with his wife and four children. The family hoped to move to Germany at some point, where Haqyar's brother-in-law lived. When we met a few years ago, he and his son had been studying German for years so they could hit the ground running when they arrived.

Then one day insurgent militants from the Taliban movement captured and brutally murdered one of Haqyar's colleagues. Terrified that he'd be next, Haqyar and his wife quickly found a buyer for their house, selling it in two days for a quarter of what they'd paid for it. They packed up their things, including several of Haqyar's German-language textbooks, which they'd need when they arrived in Germany, rounded up their four children, and left. They traveled over the mountains into Pakistan, then into Iran. There hadn't been any time to obtain official documents. When police officers sought them out, these upstanding souls ran and hid. At one point, Haqyar's wife, who struggled with a thyroid condition, went into shock and Haqyar had to carry her on his back. Later one of his sons became so dehydrated that he nearly died.

Finally the family reached Turkey, where smugglers would provide, for a hefty fee, a seat on an inflatable dinghy to cross the

Aegean Sea. It was a tantalizingly short journey: just a few miles of water separated Turkey and the Asian mainland from the Greek island of Lesbos and the rest of Europe. But while the narrow sea between Turkey and Lesbos ran shallow—during the last ice age, when sea levels were lower, it had been dry land—this migrant route could be treacherous. Many of those who tried it could not swim, and few of the smugglers equipped their boats with food, water, or safety gear. Sometimes the smugglers forced their beholden passengers into the dark, fetid spaces below deck, where toxic compounds burned their clothing and skin.

Haqyar and his family boarded one such precarious vessel. As it made its way across the waves, its engine abruptly died. The vessel drifted to and fro in the currents. Haqyar felt certain he'd drown with his children, as many others already had, their bodies washing up on the beaches of charming seaside resorts across the Greek islands. Waiters and café owners who worked along the Lesbos coast had seen such things. A photographer had captured an image, once, of the lifeless body of a three-year-old child face-down, half buried in the sand, waves gently lapping at his unmoving feet, briefly capturing the world's attention.

Haqyar and his family did not suffer that fate. In the end, they made it across the sea. Haqyar's only casualties were several of the family's precious German-language textbooks, which they'd lugged over two thousand miles from Afghanistan across mountain ranges and international borders in preparation for their new lives in Germany. The Aegean's waters had soaked into their pages, rendering them sodden and unreadable.

Haqyar discarded the ruined books on a pile of rubble left behind by the hundreds of thousands of others who traveled this route, shedding their personal items on the shores of Lesbos so they could continue their journeys west and north less encumbered. The piles grew to the height of small mountains and ridges, their primary shade bright orange from the migrants' discarded lifejackets. They glowed like beacons.

~

One of the most deeply carved migrant tracks leads out of an unlikely corner of the world, a tiny swath of land along the Red Sea on the eastern coast of Africa. In the Middle Ages it was known simply as Medri Bahra ("sea land"), later taking its name, Eritrea, from *Erythra Thalassa*, ancient Greek for "Red Sea." For decades the country's cruel, autocratic leaders forced much of its population to serve its military, burying those who dissented in secret underground prisons. Every month five thousand people from this funnel-shaped country pick up and leave, the UN estimated in 2015, traveling farther and more frequently than almost any other group of migrants.

Mariam has a watchful way about her, with deep-set eyes and a serious expression that suddenly breaks into girlish grins. She crept out of her parents' house in rural Eritrea, leaving behind her family and their small stable of livestock, at seven A.M. one morning. She'd told them of her plans, earlier. Her mother had begged her not to go, but she did anyway, she told me matter-of-factly. For nearly twenty-four hours, Mariam walked over lush mountains to the border with Ethiopia, dodging soldiers and their shoot-to-kill orders, in the first stage in what would become a nearly decade-long multinational migration. She was fourteen years old.

By leaving Eritrea, Mariam joined one of the most expansive and proportionally massive migrations in the world, its path sending out long, curling tendrils in all directions. Mariam went to Ethiopia first. Sophia, leaving behind her three-year-old daughter with her parents in the capital city of Asmara, paid a smuggler to take her by car northward to Sudan, then to Cairo.

Many others from Eritrea join the treacherous track that Ghulam Haqyar traveled, across the Aegean Sea into Europe. Some of the most intrepid make their way across the Atlantic, in hopes of reaching North America. To get there, they first must traverse an uncharted, lawless jungle in Central America.

~

Because of the difficulty of reaching the United States and Canada directly, many migrants fly to countries in South America first, and from there make the journey to the U.S. border over land. That means crossing through the delicate squiggle of land that connects the two continents, in Panama.

Ever since it rose out of the sea a few million years ago, the S-shaped isthmus has been a thoroughfare for migrants of all kinds, creating the first land bridge between creatures long separated by the waves. Biologists call the dramatic mixing and reordering that followed the Great American Interchange. North American deer, camels, rabbits, and raccoons headed south to explore and settle in warmer climes. They passed, en route, monkeys, armadillos, and opossums heading north. Those first border crossers transformed ecosystems on both sides of the boundary, sculpting the unique landscapes they're each famous for today.

Today the Panama Canal cuts through its middle, allowing ships to pass from the Atlantic to the Pacific in a few dozen miles, rather than detouring around the whole of South America to make the passage, a nearly eight-thousand-mile journey. Much of the country is crisscrossed by roads and highways, too. There's a road that runs direct from glitzy Panama City, on the country's Pacific coast, to run-down Colón, on its Caribbean side. I drove the distance, in my small white rental car, in about an hour. If I'd wanted to, I could have taken a similar road for most of the length of the country. One runs, from east to west, right up to the border with Costa Rica.

But on Panama's far eastern edge, near the border with Colombia, the roads abruptly end. There's a wide swath of untouched jungle, mountains, and swamps dripping with thick vegetation. Venomous snakes, prowling jaguars, and a maze of unmarked, mosquito-plagued trails lie within. The sultry tropical wilderness extends the entire width of the isthmus and spills over into Colombia. Because it forms the sole break in the nineteen-thousand-mile Pan-American Highway, which starts in Prudhoe Bay, Alaska, and ends in Ushuaia, Argentina, on the southernmost tip of South America, it's called the Darién Gap.

Navigating through it by vehicle is nearly impossible. Expeditionists have tried. One of the first attempts, in 1959, enlisted eight mountaineers, four crewmen, and two custom-equipped Land Rovers. After 180 river crossings, the construction of 125 log bridges, three automobile rollovers, and several bouts of malaria, the intrepid explorers of the Trans-Darién Expedition pierced the gap. The journey of sixty-six miles took them four and a half months.

A faster route is by foot and by boat, which is how today's migrants travel across the Darién Gap. They come from a wide range of countries, from Eritrea, Pakistan, and Cuba. I met several who'd arrived there from Haiti, having hopscotched through Brazil, Venezuela, and other countries in South America, en route to North America.

Thickset thirty-year-old Jean-Pierre was one. He delivers his sharp, critical observations about human behavior, in French, Spanish, and Kreyol, in a low, bitter growl. He trained as an accountant in Venezuela, but his identity, first and foremost, is as a socialist and a writer, and he sports the de rigueur goatee that proves it. He arrived at the edge of the Darién with his wife and seven-year-old son a few years ago, gathering with about one hundred other migrants at the port town of Turbo, Colombia. There, for a fee, a local boat owner would load them into some of their cargo boats for the three-hour boat ride to the Darién jungle. According to a reporter who'd witnessed migrants climbing onto Darién-bound boats in Turbo, few came prepared for the wilderness expedition that awaited them. A decent outfitter would require participants to bring, at the very least, medical kits, emergency communication devices, water filters, insecticide-treated apparel, sturdy boots, and rain gear for an expedition through this kind of wilderness. The gathered migrants in Turbo wore flip-flops. Many, like Jean-Pierre, carried small children in their arms.

By the time Jean-Pierre and his family stepped off the boat, their party had thinned significantly. Several of the boats, being overloaded and not designed to ferry people, had capsized along the way. As the unlucky flailed in Turbo's murky waters, the survivors ducked into the jungle. "The path was very narrow," remembered a young man named Mackenson, also from Haiti, who had taken the same route as Jean-Pierre. "You can't even get a horse through. People broke their legs on the trail and had to be left behind, probably to die." They walked for days. Some of Jean-Pierre's group fell from the narrow trails off the side of cliffs and into Darién's raging rivers, which swiftly swept them away. Others, straggling behind, were attacked by the drug smugglers and bandits who use the Darién's uncharted wilds for cover. At night Jean-Pierre's family slept uneasily, warding off snakes and listening to the sounds of unseen animals skulking nearby. Many migrants had resorted to drinking river water, but Jean-Pierre would not take that chance. At one low point during the journey, he, his wife, and their son drank their own urine.

After six days, they emerged out of the jungle, into a clearing not far from the road. The hundred or so others with whom they'd left Colombia had dwindled to just over fifteen. Jean-Pierre snapped a photo of the scene. Most of the frame is taken up by his wife, her back to the camera. She slouches, with her hands on her hips, in the universal posture of weariness. Her tattered top, peacock blue with white sleeves, hangs off her body in three-inch strips, exposing the dusty black bra she wears underneath. Her dark jeans are caked with mud. There are twigs in her short-cropped hair. "It was very cruel, my friend," Jean-Pierre tells me, recalling his days in the Darién. "Whenever my son thinks about it, he cries."

Jean-Pierre's family sheltered in tents for a few days in Panama, recovering and making arrangements for the next stage of their journey. Their track did not end in Panama. It continued on, snaking through over half a dozen countries and thousands of kilometers, which they'd cross on buses, on trains, and on foot, toward its final terminus: the line that separates the United States from Mexico, the most-crossed international border in the world.

~

From where I stand in the grassy expanses of the San Miguel Mountains, butterflies flitting around my feet, that border is about ten miles away, invisibly slicing through the valley between the mountains.

As I descend toward it, outlet malls, chain restaurants, and parking lots appear here and there at first, growing increasingly dense. Finally, within a few hundred yards of the border, the labyrinthine ramps and roadways and indeterminate concrete buildings seem to enclose upon themselves in a snarl, with various obscure configurations of gates and fences. Roads and freeways converge and coil, overhung with ominous signs. GUNS ILLEGAL IN MEXICO, says one; NO RETURN TO USA reads another.

One of the butterfly experts I've met grew up nearby. He remembers crossing the border as easily as a butterfly might, freely going to and fro for a fishing trip or to pick up a few lobsters for dinner. Jaguars, bighorn sheep, ocelots, bobcats, wolves, and bears regularly passed through the borderlands, seeking breeding grounds in the south and refuge from the tropical heat in the north. Birds and butterflies flew back and forth on their annual migrations, filling the skies. Today passing through the official border crossing can take hours, and it is not hard to see why. A stream of cars is backed up for miles.

Instead of joining them, I decide to park the car and walk across. Even that seems daunting. To do it, one must enter a maze-like concrete monolith, enclosed by gates and ringed by an obscure series of ramps. It reminds me of the kind of vast, multilevel parking garage that I try to avoid because I almost always end up driving in circles. The entry and exit are not easily detectable, but I successfully find the gate, and after wandering through covered walkways, up and down staircases, and across more gates, I enter a cavernous hallway where my papers are to be inspected and my bags screened. There are several booths where guards can examine documents, and stations of security screening equipment with conveyor belts.

It's empty. Nobody is there.

*Do I call someone?* I wonder. *Is there some dog-eared sign-in sheet fastened to a clipboard somewhere?* There are no signs offering any advice. Feeling discomfortingly illicit, I keep walking across the international border. Within moments, I can see the maze of shacks and high-rises amid the hills of Tijuana.

The northward flow of traffic, of course, is heavily regulated. Official border crossings—there are forty-eight, including nine in California, twelve in Arizona, and twenty-nine in Texas—dot the two-thousand-mile border between the United States and Mexico.

They process the 350 million people who cross every year. Over 150 checkpoints, situated miles beyond the crossings, spool out like a fishing net on a trawl, to capture migrants who might have slipped past the official crossings.

I pass through one, in South Texas. The signs that precede it, warning of K-9 units and federal agents, raise my blood pressure a tick despite the blue U.S. passport securely tucked in the backpack beside me. I can only imagine the dizzying spike that people such as Jean-Pierre and his family would have experienced, newly emerged from the Darién jungles, presenting themselves for inspection at one such station, in hopes of convincing officials they are worthy of passage.

Many, wary of the demand for documents, choose other routes.

In South Texas, desolate two-lane roads are the sole veins through the miles of desiccated ranchlands that line the border. Migrants heading north who prefer to avoid the checkpoints must walk through this intimidating landscape instead. Beyond the barbed wire that encloses the ranches, the sun scorches prickly vegetation. I can see the salty white imprints of shallow lakes, now dried to puddles, around which a few animals scrape sustenance: a knot of horses, a few cows. They stand silently on the parched white sand that surrounds the flat disks of stagnant water. By the side of the road, wild hogs, black and round, plunge their snouts into the crunchy bleached grass, and a gang of vultures picks at roadkill.

It takes days to cross these uninhabited, parched lands. Young, strong Cesar Cuevas told me he spent four days walking through the desert to make it north to the United States. He came prepared, carrying four gallons of water, dried meat, and tortillas. He was so good at it that the local traffickers known as "coyotes" wanted to hire him as a guide. For most others, just carrying sufficient water is tricky. The required volume—a gallon per day per person—can quickly add up to thirty pounds or more. Those who don't carry enough with them must make do with the grimy water tanks that the ranchers set out for their cattle, or the blue bins that human rights groups are sometimes able to fill for passing migrants, scribbling GPS coordinates on the inside of their lids. If they take the wrong track and don't encounter those water barrels, or fail to carry enough, or get left behind or lost, the desert sun will render them dehydrated within a few hours. It will kill them within days.

Don White, a tall gangly man with a bushy gray mustache, is a retired Motorola electronics expert and volunteer search-and-rescue expert. No one is paid to survey these desert lands along the South Texas border for stressed migrants who might need some help, so he volunteers to do it for the local sheriff's office. Every few months he fills his hydration backpack, pulls on his complicated multipocketed safari vest, and heads into the desert for a few days. He starts by hunting for footprints in the sand left behind by migrants heading north. They leave ghostly trails, which I saw, too, from the comfort of my own home, by zooming in close on the satellite images shot by Google Maps. White decides which ones to follow based on his sense of whether the people who left them behind are suffering from exposure, dehydration, or any other insult leveled by days of wandering in the desert. Dehydration changes the gait. He can see its effect in the patterns of the footprints.

Once he finds a track to follow, he has to move fast. The desert is unforgiving to dawdlers. Once the sheriff's office had received a call from a woman in Guatemala who explained that her nephew had been abandoned by smugglers near the South Texas border. All she knew was that it was somewhere near a salt lake. Ten days later White was camped out by that very lake, but he got there too late. When the wind shifted, he caught a whiff of decomposing flesh, which led him to the nephew's body and the Bible neatly tucked into the young man's back pocket.

~

A few years ago a robotics professor plotted fifteen years of refugee movements on an animated map. You can play it slowly, over the course of a couple of minutes, or, if you're impatient like me, rapidly over a few seconds. Each red dot on the map represents about a dozen refugees. At first, the dots are scattered across the map, unevenly. As the animation starts, they begin to move. Soon the red dots fuse, forming thin red lines that skitter from one part of the map to another. As more people join the journey, the thin tracks thicken, split, and radiate, creating an intricate lattice between the continents and across the oceans.

Over the last few years, biologists at the Max Planck Society created a similar video using data from eight thousand individual animals, fitted with GPS devices, as they roamed the planet. The visual effect of these collective journeys is mesmerizing. The migrant tracks move across deserts, up and down the coasts of continents, around islands in the Pacific, across oceans, and into the Arctic. Eventually they encase the planet in a delicate filigree of intertwined threads. They are everywhere.

And yet in our everyday lives, ensconced in airtight homes built on concrete foundations, we experience the landscape around us as essentially stable. Day after day I see the same faces in the grocery store aisles and wave at the same parents dropping their kids off at the school bus stop. The same scruffy squirrel runs along the top of the fence by my driveway, and the same weeds sprout out of the cracks of my front walkway. It is easy to be lulled into a sense of overwhelming sedentariness, in which the newcomer, the migrant, the intruder is the exception.

But life is on the move, today as in the past. For centuries, we've suppressed the fact of the migration instinct, demonizing it as a harbinger of terror. We've constructed a story about our past, our bodies, and the natural world in which migration is the anomaly. It's an illusion. And once it falls, the entire world shifts.

2

# PANIC

For most of my childhood, the policies and practices that
threatened global peace and security had little to do with
people moving across borders. They revolved almost entirely
around the decades-long power struggle between Washington,
D.C., and the Kremlin in Moscow.

Around the time I graduated from college, the entire edifice of
the Cold War abruptly dissolved into nothingness. In late 1989
Soviet-aligned officials in East Germany announced that the Berlin
Wall—an eighty-seven-mile-long wall encircling West Berlin and
one of the most potent symbols of the Cold War—would be torn
down. We watched on television, the night the news came out,
as thousands of ecstatic young people stormed the wall en masse
for an impromptu, all-night dance party atop it. A few months
later there was dancing in the streets again when the president of
South Africa released the revolutionary leader Nelson Mandela
from a twenty-seven-year imprisonment, ushering in the end of
the harsh system of racial segregation known as apartheid.

New graduates like me felt a deep sense of relief. The world
seemed immeasurably safer without two superpowers loudly

threatening nuclear holocaust. But soon a new global bogeyman emerged, one even more chaotic and disruptive than nuclear missiles.

The national security expert Robert D. Kaplan described it in a 1994 *Atlantic* magazine article called "The Coming Anarchy."

The magnetic poles of the United States and the Soviet Union, he explained, had held a number of destabilizing forces in suspension. Nobody had noticed, because we'd been so preoccupied with the stockpiles of missiles and the creepy binational taunting. Now, with those two poles deactivated, suppressed elements would be unleashed. Instead of improving the prospects for peace and security, the end of the Cold War would do just the opposite.

The problem: people would start to move.

As deserts spread and forests were felled, Kaplan wrote, masses of desperate, impoverished people would be forced to migrate into overburdened cities. With no great power regimes to prop up weak states, the tumult caused by migrants would result in social breakdown and "criminal anarchy." There'd be bloody conflicts. Deadly diseases would rage. Already, across West Africa, he said, young men moved in "hordes," like "loose molecules in a very unstable social fluid" on the verge of ignition. Others would soon follow. A new era of migration, he wrote, would create "the core foreign-policy challenge from which most others will ultimately emanate."

The idea of migrants as a national security threat, rushing over the land like a tsunami, captured the imagination. Kaplan's article "became required reading among senior staff in the Clinton administration," writes the geographer Robert McLeman.

National security and foreign policy experts started issuing their own reports and white papers on the threat posed by newly liberated climate-driven migrants. There'd likely be 50 million on the move by 2020, experts at the United Nations University projected.

Two hundred million by 2050, the environmental security analyst Norman Myers announced. One billion! the NGO Christian Aid projected. People moving around, in their telling, was an exceptional and future threat, "one of the foremost human crises of our times," as Myers put it.

In fact, as any migration expert could have shown, migration was just the opposite: an unexceptional ongoing reality. And while environmental changes shaped its dynamics, they didn't do so in a predictably simple way.

Migration experts had teased apart complex and counterintuitive relationships between movement and climate. They'd found that dissipating water supplies could sometimes lead not to conflict but to cross-border cooperation, which in turn could lead to less migration. During the second half of the twentieth century, for example, water scarcity had led to nearly three hundred international water agreements to cooperatively manage water sources, including between perennial enemies India and Pakistan, their agreement surviving three wars.

They'd found that the converse of the presumption that deforestation displaced people did not hold true. In the Dominican Republic, for example, restoration of forests had triggered a migrant flow, as the regreened landscape expanded the tourism industry and attracted flocks of new workers. They'd also found that sea-level rise would not automatically displace people who lived along the coast at any easy-to-calculate scale or pace. Quickly rising and receding floods might lead only to brief, short-distance migrations. Permanent and long-distance migrations would more likely follow from gradual climatic changes.

The national security experts sounding alarms about a future army of migrants took little of this nuance into consideration. They presumed that migration proceeded in response to climate stresses as a "simple stimulus-response process," as McLeman put it, "where one unit of climate change . . . triggers a corresponding additional unit of migration." They presumed that climate-driven migration would occur en masse, and with disruptive and uncontrollable effects. Water scarcity would arise, followed by conflict, which would be followed by migration. By multiplying the number of people living in those places where environmental disruptions were predicted to occur, they calculated the size of the chaotic migrations to come. The number of people who'd become migrants due to deforestation equaled the number of people who lived in places where forests were cut down. The number of people who'd migrate due to sea-level rise equaled the number of people who lived in areas predicted to be inundated by the waves. The political context, personal choices, geographic quirks, and technological possibilities that would determine such outcomes played little role.

The idea of migration as a national security threat seeped into the public's attention and incorporated itself into the world's foremost international security organizations. In 2009 the television journalist Bob Woodruff hosted a two-hour prime-time special on ABC. The special, *Earth 2100*, depicted a future world in which climate change triggers a deadly plague, which kills half the human population, and then a wave of border crossers from Mexico, which leads to the collapse of civilization. Nearly 4 million viewers tuned in to watch.

Meanwhile, in the cavernous halls of the UN Security Council, where officials debated the use of armed forces to secure the international order from threats such as drug trafficking, terrorism, and weapons of mass destruction, attention turned instead to the dangers posed by climate-driven migrants. By 2011 officials at the council had held two open debates on the subject.

At the time, the specter of mass migration had been an abstraction, like the hordes of zombies featured on hit television programs.

Then political and geographical circumstances conspired to create a spectacle, one in which migrants materialized in conspicuous masses, just as Kaplan and the others had warned, on Europe's southern shores.

~

One day in early March 2011, a few bored teenagers in Daraa, a dusty Syrian town decimated by years of drought and neglect, found a can of red paint.

The boys could have used the red paint to scrawl their names somewhere, or those of their sweethearts. But images of revolution dominated their television screens. Uprisings and protests against oppressive, autocratic leaders had erupted across the region. In just a handful of weeks, mass demonstrations in Tunisia and Egypt had overthrown a government and forced a dictator to resign.

Little of the revolutionary fervor of what would come to be called the Arab Spring had reached sleepy Daraa or anywhere else in Syria yet. A Facebook-organized "Day of Rage" against the Syrian leader Dr. Bashar al-Assad had fizzled, failing to draw much of a crowd. Inspired by what I imagine must have been a combination of frustration, boredom, and cheek, the teenagers carried the can of red paint to the local school and brushed its contents into a three-word warning dripping from the wall: "Your turn, doctor."

It may have seemed harmless enough at the time.

Unexpectedly, the enraged Assad regime detained and tortured the teens. As the news seeped out, demonstrations erupted across the country, drawing yet more brutality from Assad. Soon Syria descended into a bloody civil war. In time hundreds of thousands would perish.

The boys' small act of resistance ended up sparking one of the most brutal civil wars in recent history.

The war in Syria unleashed a mass exodus. People streamed out of the country in all directions, like water from a sieve. Hundreds of thousands sought refuge in Iraq and Jordan. Over a million ended up in nearby Lebanon. Nearly 2 million headed toward Turkey, en route to Europe.

At the same time, the Arab Spring opened another valve for migration into Europe. While Libya's autocratic leader Muammar Gaddafi had been in power, few migrants had been able to successfully migrate through the country to get to Europe. But during the Arab Spring, a U.S.-led military alliance helped topple and murder that leader, and with him, the security infrastructure that had once prevented migrants from passing through the country. As migrants from all over sub-Saharan Africa started to converge in Libya to make the passage to Europe, a lucrative smuggling trade sprang up to help them.

The flow of migrants from Syria into Europe, joined by the second flow of migrants newly able to pass through Libya, soon turned into an international spectacle, dominating headlines in Europe and North America. It wasn't necessarily the scale of the migration that captivated. More migrants moved between different countries within Africa and Asia than into Europe. But the migrations into Europe, unlike those across jungles and mountains elsewhere, converged from various directions in a single high-profile and especially picturesque choke point: the Mediterranean Sea.

∼

Named after the Latin for "middle," *medius*, and "land," *terra*, the Mediterranean Sea is squeezed in between land masses, with Europe to its north, Africa to its south, and Asia to its east. It's a peculiarly accessible body of water, thousands of miles long but just a few miles wide at its narrowest point. Nearly twenty different countries claim a bit of its coastline. When, over the course of 2015, over a million people pushed into its waters aboard overcrowded, rickety vessels—more than 850,000 from the coast of Turkey and another 180,000 from the coast of Libya—the bedraggled armada couldn't help but capture attention.

Photographers shot images of their wide-bellied wooden boats and flimsy rafts, some caught in the midst of capsizing, their passengers clinging helplessly to gunwales or splashing in the sparkling sea. They captured pictures of the lifeless bodies of the drowned, washed up on the beaches of Greek islands. Filmmakers, artists, and celebrities of all ilk descended on the Greek islands where the migrants' boats landed, recording video of themselves helping unload cold frightened migrants off boats and warming them up with cups of tea, including the actors Susan Sarandon and Angelina Jolie, the activist artist Ai Weiwei, and Pope Francis, the head of the Catholic Church. Thousands of new migrants arrived every day, fanning out into the rest of Europe any way they could, on foot, by bus, and by train.

Press reports immediately dubbed the arrival of the newcomers a "migrant crisis," describing a "migrant invasion" in which migrants "stormed" ports and ferries, taking whole cities "hostage." According to one analysis of press coverage in Europe at the time, nearly two-thirds of articles "strongly emphasized" the various negative consequences the migrants would effect—even in the early days when no such impacts had actually yet occurred—and the same proportion could think of no positive consequences of their arrival, neither real nor projected. Reporters described the newcomers themselves in only the most cursory way, rarely referring to them as full individuals with names, ages, genders, and professions. Most mentioned only one characteristic: their foreign nationality.

The possibility that Europe, with its total population of over 500 million, could absorb another million people went mostly unexplored. In fact, countries such as Greece and Hungary had plenty of accommodations and jobs to offer newcomers. In Athens, three hundred thousand residential properties stood vacant. In Hungary, a critical labor shortage meant employers couldn't find sufficient workers to fill vacant posts.

But for many observers, the newly conspicuous spectacle of mass migration appeared ominous. They saw an army of robotic migrants, full of disruptive and destructive potential.

~

By 2015 over a million people from Syria, Afghanistan, and elsewhere had found their way into Europe, primarily Germany but also Sweden and elsewhere. In their wake, a wave of politicians promising harsh new measures against migrants swept into power across Europe and the United States. U.S. voters elected Donald Trump, an unlikely populist who derided people from Mexico as rapists and criminals and led crowds in chants to "build the wall" that would stymie their movements. The people of Britain voted to leave the European Union and its open borders altogether. Political parties that vowed to fight the invasion of foreigners, refuse entry to even a single refugee, and intern refugees in camps won unprecedented numbers of seats in European parliaments, capturing the majority of seats in Poland, their first parliamentary seats in Germany, and joining the governing coalition in Austria. A politician who refused to admit any refugees whatsoever became prime minister in the Czech Republic. Another, whose party proposed expelling all migrants, became prime minister of Italy.

Government agencies once dedicated to welcoming immigrants repurposed themselves as defenders against them. The U.S.

Citizenship and Immigration Services, which had described its purpose as fulfilling "America's promise as a nation of immigrants," revised its mission statement in early 2018, excising those words. Its new commitment would be toward "securing the homeland." The message from inside the newly fortified borders of Europe was equally clear. The European Union head Donald Tusk, whose organization had been founded on the principle of open borders, spoke plainly. "Wherever you are from," he said, "do not come to Europe."

As antimigrant politicians climbed into power, reinforcing the urgency and necessity of the antimigrant policies they touted became a political necessity. Like any regime, they and their supporters would have to continuously justify their political stances. Emphasizing the mayhem caused by migrants would be key to that project.

The effects that experts predicted—crime waves, epidemics, and economic catastrophe—were not subtle. Given the scale of the influx of newcomers, showcasing evidence of the chaos they caused should have been easy.

~

During the first days of January 2016, scores of women showed up at police stations in cities across Germany to file complaints about what had happened on New Year's Eve. As they'd been making their way to trains and homes after the New Year celebrations, they said, they'd been surrounded, groped, robbed, and sexually assaulted. The attackers, from what the women could tell from their clothes and accents, had been newly arrived migrants from Arab and North African countries.

Media outlets featured stories suggesting the newcomers had a special appetite for raping local women. In Germany, a magazine cover story featured an image of a white female body covered in muddy handprints. "Women complain of sex attacks by migrants," the caption read. "Are we tolerant or are we blind?" Another ran an interview with a psychologist about the "mentality" of Arab men, illustrated with an image of a black hand reaching between a pair of white legs. In the Netherlands, a newspaper printed a reproduction of a painting called *The Slave Market*, in which Arab men disrobed white women before selling them as sex slaves. In Poland, a magazine ran a cover story on "The Islamic Rape of Europe," featuring an image of black and brown hands tearing a European-flag-printed dress off of a blond woman's body.

That spring the German Interior Ministry released a report showing that the country had experienced 402,000 excess crimes since admitting its latest wave of migrants, a breathtaking statistic featured prominently in newspaper reports around the world. In one particularly inflammatory episode, a mob of migrants had been captured on video chanting and celebrating after setting one of Germany's oldest churches on fire.

News of the migrant-driven crime wave in Germany swept across the Atlantic. In the United States, popular right-wing news outlets such as Breitbart ran stories on the "New Year Rape Horror" committed by "rape-fugees." "Crime in Germany is way up," Trump tweeted to his millions of followers. "Big mistake made all over Europe in allowing millions of people in who have so strongly and violently changed their culture!"

News from Sweden a few months later appeared to suggest that the criminal anarchy taking hold in Germany had ignited there, too. Sweden had accepted more migrants per capita than any other country in Europe. A documentary filmmaker from Los Angeles named Ami Horowitz visited Sweden to report on the situation. He found that reports of rape in the country had skyrocketed. Sweden, once known for stylish furniture and saunas, "is now the rape capital of Europe," he reported.

Entire neighborhoods had been subsumed by the new migrants. A leafy suburb of Stockholm called Rinkeby had become a "completely Islamic area," Horowitz said. Local police officers told him that Rinkeby, like many of Sweden's new migrant-dominated neighborhoods, had become so lawless that they feared to enter. They called it a "no-go" zone. Gunshots rang out daily. Armed twelve-year-olds roamed the streets. A band of young migrants surrounded and attacked a *60 Minutes* film crew from Australia that had attempted to film in Rinkeby. Horowitz had seen the harrowing footage they'd shot himself.

Horowitz's documentary on the migrant crisis in Sweden, *Stockholm Syndrome*, aired on Fox News's website in the fall of 2016. A few months later the conservative commentator Tucker Carlson interviewed Horowitz on his prime-time current affairs show, watched by nearly 3 million viewers. Sweden, Horowitz explained, was under assault "because of the open door policy to Islamic immigration." The next day newly elected president Donald Trump mentioned Horowitz's findings to nine thousand fans at a rally in Melbourne, Florida. "Look at what's happening last night in Sweden," he called out to the raucous crowd. "Sweden, who would believe this?"

Within days, right-wing media outlets broadcasted news about the crime wave in Sweden across the nation. Commentators such as the right-wing journalist Bill O'Reilly, whose show reached over 2 million viewers every night, featured interviews with experts such as a Swedish defense and national security adviser who confirmed Horowitz's alarming findings.

While government reports and flashy news stories depicting migrants as criminal piled up, an equal volume of critiques poking holes in their underlying logic accumulated alongside, like a ghostly doppelgänger.

In the summer of 2017, an NPR reporter examined the reports of rising crime in Germany. The attacks on New Year's Eve had transpired, he found, but may not have been exceptional. Sexual violence in Germany, as elsewhere, constituted an ongoing crisis, with over seven thousand rapes and sexual assaults reported every year, affecting more than a third of women in the country. Many more assaults, experts said, went unreported. And the country's annual New Year's Eve celebrations provided ample cover for all manner of criminality, as one BBC correspondent explained, turning the streets of German cities into a cross between a wild drunken party and a riot. "The drunkenness on the streets is of a level I don't think you'd really see in the States," he told an NPR reporter. The difference in 2015 may have been that the perpetrators weren't the familiar sexual predators long known to be roaming the country.

There hadn't been any concomitant crime wave. The 402,000 "excess" crimes consisted entirely of the "crime" of crossing the border without prior permission, a transgression that by definition could have been committed only by newly arrived migrants, as a closer reading of German government reports stated. Extracting those violations from the data revealed that crime rates in Germany had remained pretty much the same the year after thousands of migrants had arrived as the year before. By 2018 crime in Germany reached its lowest rate in thirty years.

That same NPR reporter found no evidence that the newly arrived migrants intended to destroy German institutions either. Migrants hadn't set any churches afire. The Christian church that had supposedly been burned down by gleeful Syrian migrants had not been purposely set on fire, as the inflammatory video captured of the conflagration suggested. What had happened was considerably more mundane, as law enforcement officials and local newspaper reporters clarified. Syrian refugees had been celebrating a cease-fire in Syria. During the celebration, one of their fireworks had briefly ignited some netting on the scaffolding of the church. The video clip of the flames had been taken out of its context.

There hadn't been any crime wave in Sweden. When journalists followed up on Horowitz's claims, they found no evidence for any of them. The Swedish expert interviewed on Fox News, who'd been presented as a "national security adviser" in the country, had in fact "not lived in Sweden for a very long time," a professor at Swedish Defence University told the *Washington Post*. "And no-one within the Swedish security community . . . seems to know him."

Stockholm was no "rape capital." According to the Swedish crime survey, 0.06 percent of the population reported having been raped in 2015. That compared favorably to England and Wales, for example, where 0.17 percent of the population did, reporters from *Vice* found. The so-called "no-go" zones didn't exist. Two of the police officers interviewed by Horowitz in his documentary said their words had been taken out of context. "He has edited the answers," one told a reporter from *Dagens Nyheter*, one of Sweden's largest daily newspapers. "We were answering completely different questions in the interview."

Important context had been left out of Horowitz's description of the migrant attack on the Australian film crew as well. There had been a scuffle between young migrants in Rinkeby and the film crew, local police officials told a reporter from Sweden's public radio service, but the crew had not been the neutral observers that Horowitz's presentation implied. They'd been working with a website called Avpixlat, which international and local media described as a racist, anti-immigrant hate site. There'd been no damages or injuries, and the police ended up dropping their investigation into the brouhaha.

In meadows and forests, one can sometimes find an unusual species of mushroom called *Calvatia gigantea*. *Calvatia* don't have the typical umbrella-shaped form of other mushrooms, with stalks topped by drooping spore-lined caps. Instead, they grow into massive white spheres, as large as soccer balls. Their spores build up invisibly inside them. At maturity, so-called "puffball" mushrooms like *Calvatia* become so intensely packed with spores that any minor impact—even a drop of rain—can puncture their exterior. If you poke one with a stick, or give it a little kick, a smoky cloud of spores will explode from the interior, leaving behind nothing but an empty, crinkled shell.

The narratives used to justify antimigrant policies turned out to be similarly bloated and hollow. With even the lightest scratch to their surfaces, they dematerialized into a cloud of smoke.

∾

Although the number of unauthorized immigrants entering and living in the United States had been falling since 2007, government reports and antimigrant politicians portrayed the criminality of migrants in the United States and along its borders as similarly emboldened, as if strengthened by some invisible current from across the Atlantic. Along the U.S.-Mexico border, government reports under President Donald Trump showed that attacks on Border Patrol agents spiked, increasing by 20 percent in 2016, then by over 70 percent in 2017. The men and women who guarded the border suffered the highest rate of assaults of any group of federal law officers, the chief of the U.S. Border Patrol told lawmakers in testimony to Congress. And it was getting worse. "Year-to-date we're seeing an increase in assaults up to 200 percent from the previous year to date," he said.

In the fall of 2017, the bloodied bodies of two Border Patrol agents were discovered at the bottom of an eight-foot concrete culvert along the border in West Texas. "There's a high likelihood this was an assault on the agents," a Border Patrol officer told reporters. That assault, a Fox News television host added, had been "most gruesome." The poor officers had been "brutally beaten," President Trump informed his Twitter followers. In fact, figuring out what had happened had required some sleuthing: one of the agents died soon after being rushed to the hospital, and the other suffered a brain injury that resulted in confusion and memory loss, so neither officer could tell anyone what had happened to them. But since the survivor had suffered a blunt force trauma to the head, and he and his partner had been found surrounded by rocks, Border Patrol agents surmised that there must have been some kind of ambush. A gang of migrants had surrounded the agents and pounded their heads in with the telltale rocks. Such a scenario neatly explained the available evidence, Border Patrol officials said.

The Texas governor, presuming the scenario to be true, offered a reward to anyone who could help authorities catch and punish the culprits who had committed such a barbarous attack. The assault epitomized the threat the nation faced from an "unsecure border" that allowed migrants with criminal intents to pour in from the south, the Texas senator Ted Cruz explained.

New analyses purported to reveal the security threat that migrants posed even when they resided well inside the nation's borders. The Department of Homeland Security issued a report in early 2018 showing that three-quarters of defendants convicted on international terrorism charges between 2011 and 2016 had been born outside the United States. The report, the attorney general claimed in a statement announcing it, "reveals an indisput-able sobering reality—our immigration system has undermined our national security and public safety." Migrant crime had become such a crisis that the president created a special government office

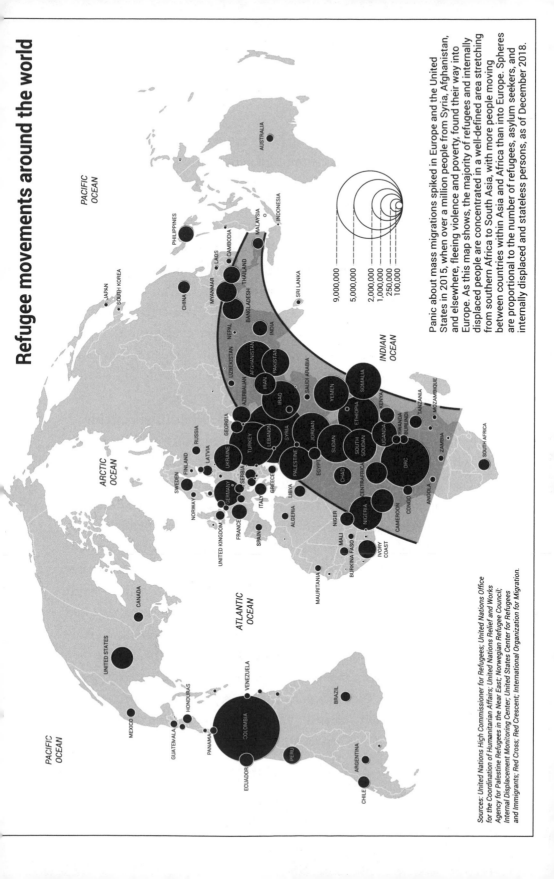

# Refugee movements around the world

Panic about mass migrations spiked in Europe and the United States in 2015, when over a million people from Syria, Afghanistan, and elsewhere, fleeing violence and poverty, found their way into Europe. As this map shows, the majority of refugees and internally displaced people are concentrated in a well-defined area stretching from southern Africa to South Asia, with more people moving between countries within Asia and Africa than into Europe. Spheres are proportional to the number of refugees, asylum seekers, and internally displaced and stateless persons, as of December 2018.

9,000,000
5,000,000
2,000,000
1,000,000
250,000
100,000

Sources: United Nations High Commissioner for Refugees; United Nations Office for the Coordination of Humanitarian Affairs; United Nations Relief and Works Agency for Palestine Refugees in the Near East; Norwegian Refugee Council; Internal Displacement Monitoring Center; United States Center for Refugees and Immigrants; Red Cross; Red Crescent; International Organization for Migration.

entirely dedicated to serving the victims of crimes committed by migrants.

Politicians and right-wing news media highlighted case after case in which migrants committed acts of brutality. In one notorious case near my own neighborhood outside Baltimore, two undocumented migrants reportedly gang-raped a fourteen-year-old girl in a high school bathroom. Rather than attributing such crimes to the ongoing epidemic of violence against women and girls, politicians argued that allowing migrants into the country was the problem. "We need to know who is within our borders and who does not belong," one local councilman wrote to his constituents in response to the crime. "Immigration pays its toll on our people if it's not done legally," explained a White House spokesman, referencing the rape. "This is another example."

As social panic about migrant-driven criminal anarchy spread, experts and officials started to conflate migrants with criminals, whether they'd violated criminal laws or not. Administration officials described efforts to rid the country of unwanted migrants as tantamount to fighting crime itself. The attorney general scolded a northern California mayor who had thwarted federal immigration officials' efforts to root out undocumented migrants for allowing "wanted criminals" to roam "at large." The special adviser to the director of U.S. Immigration and Customs Enforcement (ICE) said that ridding the country of criminal gangs required targeting all migrants more generally, because migrants and criminal gangs were one and the same thing. ICE officers classified migrants as gang members, whether or not any evidence suggested they were. One ICE officer brought a reporter along on a raid in which he officially marked a captured migrant as a gang member, despite having no evidence on which to base that assessment. "The purpose of classifying him as a gang member or a gang associate," he explained, was not to accurately reflect any evidence collected against him. It was "because once he goes in front of an immigration judge, we don't want him to get bail."

∾

As in Europe, the case for the migrant crime wave in the United States had been manufactured.

There was no spike in attacks against Border Patrol agents on the U.S.-Mexico border. In 2015 the Border Patrol had changed its method of counting assaults on officers. Instead of counting the number of agents assaulted, as most other experts did and as they had done in the past, officials started counting the number of agents assaulted, multiplied by the number of attackers, multiplied by the number of objects the attackers used in the assault. If a few migrants threw a couple of rocks or sticks at some Border Patrol agents, each rock and each stick thrown by each migrant counted as a separate incident.

On February 14, 2017, for example, six people threw rocks, bottles, and sticks at a group of seven Border Patrol agents. Border Patrol officials generously logged that single incident as 126 separate assaults. Their unusual new method entirely accounted for the elevated numbers of assaults on Border Patrol agents they had reported, as an investigation by the immigration reporter Debbie Nathan revealed.

Using more traditional methods of tallying assaults, the statistics showed that Border Patrol agents did not experience the highest assault rate among law enforcement officers. They experienced the lowest. The death rate among Border Patrol agents was about one-third that of the nation's law enforcement officers who policed residents.

The Border Patrol agents who had been found bloodied at the bottom of the culvert in West Texas hadn't been ambushed by migrants, FBI officials and the local sheriff told the *Washington Post*. After more than two months of investigation, the FBI found no evidence of any "gruesome" beating or of any attack whatsoever. The officers had not discharged their weapons. No evidence suggested that anyone had attempted to take their weapons, either. The officers had been patrolling a difficult patch of terrain on a dark moonless night, and their bodies had been found at the bottom of the eight-foot culvert. They'd fallen down. The surviving officer, before his brain injury jumbled his memory, had even said that the two had "ran into a culvert" when he first called for help.

The Department of Justice report finding that three out of four people convicted of international terrorism had been born outside the country was accurate, as far as it went, but it didn't support the attorney general's statement, while announcing the results, that immigration "undermined our national security and public safety." That's because international terrorism accounted for only a fraction of all terrorism attacks, which included both international and domestic charges, as the investigative reporter Trevor Aaronson pointed out. Whether foreign-born people composed the majority of people convicted on *all* terrorism charges was unclear: the Justice Department maintained a list only of those convicted on *international* terrorism charges. They had no such list of people prosecuted for domestic terrorism.

The most gruesome and widely commented on anecdotes about crimes committed by undocumented migrants turned out to be similarly spectral. Investigators dropped charges against the undocumented migrants who'd been accused of gang rape in a high school bathroom, when the alleged victim's story fell apart. Neither the White House nor local politicians in Maryland, who'd held up the crime as evidence of migrants' generally suspicious proclivities, issued any corrections to the constituents they'd misled by the thousands.

~

Historians of disease have never found any systematic association between infectious disease and modern migration. But the suspicion that migrants might cause epidemics persisted nonetheless, based on what seemed to be faultless logic. Vaccination programs in the countries many of the migrants had fled were either nonexistent or had broken down. In theory, that meant that newcomers could harbor pathogens that had been controlled in the countries they'd entered, sparking deadly epidemics.

Public health researchers in Europe started examining migrant bodies in more detail to find out. They found that tuberculosis rates in Germany spiked by 30 percent in 2015, when migrants streamed into the country. In Britain, foreign-born people composed just 13 percent of the population but more than 70 percent of the tuberculosis cases and more than 60 percent of malaria cases, public health researchers reported. In Italy, researchers discovered a strange cast of microbes lurking inside the bodies of Syrian refugees, including "unusual species of bacteria and fungi rarely circulating in Italy or in other developed countries," including some that "could represent a potential dangerous pathogen . . . that could spread." In Germany, doctors discovered refugees infected with salmonella and shigella. In Switzerland, they discovered refugees with rates of antibiotic-resistant bacteria five times higher than in the resident population.

Fears of a migrant-driven epidemic flared. In Bulgaria, a 2013 study of articles about migrants found that the two most commonly appearing words were *threat* and *disease*. "Send home killer bug migrants," newspaper headlines in Britain blared. "Migrants with TB should be sent home." In Greece, right-wing vigilantes aiming to root out sickly migrants marched into hospitals and demanded that patients and doctors provide them with their residency papers. Although no outbreaks caused by migrants had yet occurred, "there are already signs of the emergence of very dangerous diseases which haven't been seen in Europe for a long time," an antimigrant politician in Poland said. "Cholera on Greek islands; dysentery in Vienna; various types of parasites, protozoans, which aren't dangerous in the organisms of these people but which could be dangerous here." Donald Trump proclaimed that migrants would ferry contagion into the country. In his unique locution, migrants turned into pathogenic germs themselves. "Tremendous infectious disease," he said, "is pouring across the border."

But the presence of microbes in migrants' bodies did not by itself signify that they posed any more or less of a health risk to others than anyone else. Scrutinizing any human body for microbes is likely to reveal a long laundry list of suspicious-sounding characters. Public health researchers exposed their presence in refugees' bodies by taking intrusive rectal swabs, but they hadn't subjected residents' bodies to the same surveillance. "If you did that to people in UK," noted one public health specialist who worked with migrants, "they would have it too."

In fact, certain high-profile groups of migrants, such as refugees who entered the United States, were among the most rigorously health-screened and vaccinated residents in the country. Their bodies likely posed less risk to others than those of the resident population. And after more than a million migrants entered Europe, the continent had seen little other than a few outbreaks of minor ailments, all of which had been quickly detected and controlled.

∾

Economists had long struggled to detect any negative economic effect migrants imposed on locals. That changed in 2015, when the Harvard economist George Borjas claimed to uncover evidence that migrants exacted a costly economic burden. Borjas analyzed the effects of a rapid influx of migrants on the labor market in Miami, finding that their arrival had had a "dramatic" and "substantial" effect on high school dropouts, whose wages declined by as much as 30 percent.

Borjas's results overturned decades of analyses by other economists. They'd used the same data—from what was known as the "Mariel boatlift," an episode during which over one hundred thousand people had boarded vessels at the port of Mariel in Cuba and fled to Miami—but had found no effect either on wages nor on employment, compared to other cities that hadn't experienced any migrant surge.

Borjas had ferreted out migrants' burden on the economy by isolating their economic impact on high school dropouts. By doing so, he had "nuked" the Mariel example as a case study of the strangely negligible economic impact of migrants, the conservative commentator Ann Coulter proclaimed to her six hundred thousand Facebook followers.

Trump's attorney general Jeff Sessions considered Borjas "the world's perhaps most effective and knowledgeable scholar" on migrants' impact on the economy. His conclusions that migrants depressed wages "deeply influenced" Sessions, the *New York Times* reported. Citing his study, the White House adviser Stephen Miller argued that the United States should slash the number of migrants allowed into the country by half.

New government analyses detailed other economic damages supposedly wreaked by migrants. The Department of Health and Human Services reported in 2017 that refugees required more costly social services per capita than the typical U.S. resident did. Between 2011 and 2013 they had cost the U.S. economy over $55 billion, the National Academy of Sciences found. "Refugees with few skills coming from war-torn countries," a White House spokesman explained, "take more government benefits . . . and are not a net benefit to the U.S. economy."

"Immigrants," the president proclaimed in his 2017 address to Congress, cost the United States "billions."

In fact, Borjas had left out a potentially confounding factor. During the period in Miami he'd studied, the Census Bureau had changed how it counted high school dropouts, in a way that led to many more being counted in Miami than in other cities that Borjas had used for comparison, the migration expert Michael Clemens pointed out. Borjas had attributed the decrease in high school dropouts' wages to migrants, but the Census Bureau's changed methodology could have accounted for the apparent decline entirely. And the economic benefits contributed by refugees more than offset the costs they incurred in government benefits. Over the past decade, refugees in the United States had brought in $63 billion more than they'd cost, the *New York Times* and other news outlets reported. The National Academy of Sciences report had found that immigrants cost the U.S. economy $57.4 billion between 2011 and 2013, but that same report found that the children of those immigrants added a net benefit to the economy of $30.5 billion, and their grandchildren added a whopping $223.8 billion.

～

"Many people are being killed!" a local immigration expert informed a small crowd assembled in the fluorescent-lit banquet hall of the American Legion building. The expert, Jonathan Hanen, a paunchy man with a receding hairline and childlike rosy cheeks, delivered his presentation gripping the podium in both hands, his tall stooped frame lurching at a forty-five-degree angle. He'd been invited by the Republican club in my hometown to "take a muddy issue and make it clear," as the club's president put it in his introduction. And he did, distributing a dense, fourteen-page handout crammed with tables and charts, showing how "illegal aliens" committed a disproportionate number of crimes, plunging the nation into crisis. "One day after graduating, a 4.0 GPA student was run over by an illegal alien," he told the crowd. "You have these stories all over the country."

It was true that undocumented migrants were overrepresented in federal crime statistics, as Hanen's handout prominently mentioned. But that didn't support Hanen's claim that migrants committed more crimes than residents. Federal crimes represent only a fraction of crimes committed in the country, 90 percent of which appear in state and local crime statistics. While no nationwide data tracked offenders by their immigration status, social scientists had found that neither places with higher proportions of immigrants nor those with new influxes of immigrants suffered higher crime rates. Between 1990 and 2013, the number of undocumented immigrants in the United States tripled, but the rate of violent crime in the country nearly halved.

Hanen did not mention it. Like many of the immigration experts disseminating faulty data about migrants as educational fare, he was a bit of a puffball mushroom himself. He was neither an educator nor even much of an immigration expert. He had a PhD in ancient Greek philosophy and practiced what the ancients might have called "sophistry," working as a propagandist for ideological think tanks, political campaigns, and antimigrant lobby groups.

Immigration wasn't an especially pressing issue for most of the attendees that cold January evening. Early in his talk, Hanen had peered at the small crowd through his thick black-framed glasses. "Who here knows who Emma Lazarus was?" he asked, referring to the poet who'd written the famous words inscribed at the base of the Statue of Liberty welcoming the "huddled masses yearning to breathe free." The attendees squirmed, furtively glancing at one another. Most of the middle-aged professionals there had arrived straight from their offices, still wearing their sensible shoes and rumpled suits. They were more interested in chatting about a new Young Republicans club at the local high school and enjoying a cold Yuengling and a slice of pizza than in revisiting milestones in U.S. history. The journeys of their fellow human beings, across sea, desert, jungle, and mountains were as distant as a Komodo dragon in that fluorescent-lit suburban hall, with its practical low-pile wall-to-wall carpeting. No one raised a hand.

Still, the gathered club members nodded through Hanen's talk. When he proclaimed, triumphantly, that "Emma Lazarus was not elected to Congress!" to justify closing the borders, they tittered, though many presumably still did not know who she was. After the talk, they clapped politely and asked Hanen a few general questions. But even though the migrant crisis he described hadn't especially gripped them, some would no doubt take him up on his offer to deliver his presentation to other groups they belonged to, and use the handout he'd provided to make their own three-minute statements on immigration at public meetings and to their elected officials. Even if they didn't, they'd at least remember a few details, or some general impressions, which they'd take home to share with their kids and neighbors. They'd bubble up in casual remarks at the soccer field and around sports bars and family barbecues.

Seemingly neutral nuggets of information about the criminality and sickness of migrants infiltrated the cultural conversation and spread far and wide. By 2017 even residents in Homer, a town of around six thousand souls at the end of the U.S. road system in Alaska, had heard the news about the migrant crisis in Europe and prepared to gird themselves against an onslaught. "You bring in illegals, OK, by definition, they're criminals," one resident explained heatedly at a Homer city council meeting, in response to an ill-fated proposal to welcome any immigrants who might find their way to Homer. None ever had, and few were likely to, given the town's remote location. "OK, they live in the underworld. They don't have a stake in the game as we do. About the first time somebody gets raped or killed, I hope they come straight after the Homer City Council and sue!"

Scenes such as these replayed in communities across the United States and Europe. And as they did, a picture of migrants as a global threat lodged itself in the public mind. Its size ballooned. Americans and Europeans alike vastly overestimated the proportion of immigrants among them. Americans, in one study, overestimated the proportion of immigrants in the country by 200 percent. Half or more people in a range of European countries believed that newly arrived refugees made terror attacks more likely. Forty-five percent of Americans believed that immigrants worsened crime.

The president described a 2018 caravan of migrants two thousand miles from the U.S. border as an "invasion of our country," with "criminals and unknown Middle Easterners" mixed in. He sent troops to the border to repel them. The migrants, a woman in Sparta, Illinois, said at one of the president's rallies, were "a plot to destroy America, and to bring us to our knees . . . I'm not going to take it—not going to go down without a fight." Their arrival, a radio host said, would spell "the end of America as we know it."

The corrections and clarifications punctured the puffball, revealing the hollowness of its interior, but could not destroy it. The spores lifted into the air and were carried in the breeze to other locales, where they settled, took root, and sent out new shoots.

∾

In early 2018 the U.S. president gathered a few lawmakers together for a private meeting in the Oval Office to discuss the country's immigration policies. "Why are we having all these people from shithole countries come here?" he demanded of them, in comments that leaked to the press. His attention turned to one group of migrants in particular. "Why do we need more Haitians? Take them out." People from Haiti "all have AIDS," he'd grumbled some months earlier.

People had fled Haiti en masse after a devastating earthquake hit the island in 2010. The U.S. government allowed about sixty thousand Haitians to stay in the country under a program known as "temporary protective status" (TPS), which granted eighteen months of legal status to people from countries that suffered natural disasters or protracted unrest. Haitian earthquake survivors arrived in the United States on airlifts still covered in the dust from the rubble from which they'd been extracted.

But the welcome did not last. A few months after the quake, U.S. officials sent Air Force cargo planes to Haiti to broadcast the message that anyone who dared try to come to the United States would be arrested and turned back. Thousands of Haitian quake survivors, shut out of the United States, migrated to Brazil and elsewhere instead.

Then the Brazilian economy tanked. The Haitian quake survivors who had settled there, such as Jean-Pierre and his family, were set into motion once more. By late 2015 thousands had amassed on the U.S.-Mexico border, hoping to gain admission and join the earlier wave of quake survivors settled in the United States. But this time White House officials were not in a welcoming mood.

For years U.S. immigration officials had regularly renewed Haitian immigrants' temporary protective status every eighteen months. The crisis that had precipitated their need for refuge, after all, continued. Abruptly, in November 2017 the director of the U.S. Citizenship and Immigration Services L. Francis Cissna proclaimed that he'd found that Haiti had "made significant progress" in recovering from the 2010 earthquake. That meant that the country "no longer continues to meet the conditions for [TPS] designation."

The Haitians waiting on the southern border to enter the country would be summarily deported back to Haiti. Families who'd established homes and businesses—nearly half of people with TPS owned homes, and more than 80 percent participated in the labor market, compared to just over 60 percent for the rest of the U.S. population—would have to leave the country voluntarily or face deportation.

Emmanuel Louis, a lawyer from Port-au-Prince who'd arrived after the earthquake, heard the news while working as the night shift as a nusing assistant. "You laugh and you are happy and then someone says, you have to go see the office manager," he remembered. "You are happy, you think you are going to get a raise! And they say, you know what, your work permit is about to expire." His friends stopped going to work and kept their kids home from school. "They are afraid of everything," he said. "Everyone says to each other, be careful, be careful!"

Community workers across the country started advising their frightened Haitian clients to memorize the phone numbers of people they'd need to call when immigration officials collected them for deportation. Their phones would likely be confiscated. The homes and businesses owned by people like Emmanuel Louis would be lost, too, unless they started the process of transferring titles to others now. So would their tens of thousands of U.S.-born children, who'd become wards of the state when their parents were deported. Shell-shocked parents had to start preparing to transfer custody of their children to others, community workers advised.

Jean-Pierre's family barely escaped summary deportation. Although they'd been held in detention after crossing into the United States—his seven-year-old son, still having nightmares about the snakes in the Darién jungle, had even been handcuffed— they'd been allowed to leave after a week, pending a later court date to hear their claim for asylum.

Jean-Pierre had made it to Orlando when he heard that one of his friends had been deported, after being held in detention for a year. Jean-Pierre had experienced more than his share of soul-crushing trauma. He'd survived an earthquake that killed hundreds of thousands, gang violence in Haiti in which his and his relatives' lives had been threatened, and a death-defying journey to seek refuge that had required, among other things, that he drink his own urine to survive. Not to mention that he worked at Disney World, which for a committed socialist was likely a kind of agony, too. But it was the news about the deportation of his friend that broke him. He said he felt like killing himself.

$$\sim$$

If migrants were as sickly, criminal, and economically disastrous as antimigrant politicians claimed, it would have been easy for the administration to build its argument for Haitians' eviction. In fact, it struggled to concoct its case. According to emails leaked to the Associated Press, Trump administration officials had had to actively hunt for data on the number of Haitians and other immigrants with TPS who had been accused of crimes or had illicitly collected public benefits. "Find any reports of criminal activity by any individual with TPS," one immigration official had instructed her staff. "We need more than 'Haiti is really poor' stories" that supported their continued need for refuge in the United States.

In claiming that Haiti had made "significant progress," Cissna, the director of U.S. Citizenship and Immigration Services, had ignored the findings not only of agencies such as the State Department but also of his own staff. In an internal report, staffers at USCIS had found that the difficult conditions that led people to migrate from Haiti continued to persist. Food was scarce, cholera was rampant, and repeated natural disasters "severely worsened the pre-existing humanitarian situation."

According to a State Department travel advisory, political violence was rampant. "Protests, including tire burning and road blockages, are frequent and often spontaneous," the State Department warned. "Kidnapping and ransom can affect anyone," and "Haitian authorities' ability to respond to emergencies is limited and in some areas nonexistent." State Department officials had judged the security situation in Haiti to be so bad that they didn't allow embassy employees to travel there except with special permission. Even then, the State Department said, they should have plans for "quickly exiting the country if necessary."

⁓

Darrell Skinner, a heavyset Texan wearing a stiff baseball hat and oversized tinted sunglasses, hunched in his red-vinyl-upholstered seat at Dinks Cafe, a roadside diner in Del Rio, Texas, about twenty miles from the border between the United States and Mexico.

"If we don't do something about the border immediately," Skinner declared, "this country won't be in existence in fifty years." The café owner, Cheryl Howard, whose blond bob was held back from her face by reading glasses perched atop her head, agreed. "We need to keep them over there," she said conspiratorially, glancing around to make sure none of her Mexican customers could hear.

Even as the hollow core of the case against migrants lay exposed, certainty about the existential threat they posed persisted. It bubbled up from a deeper sense of violation. The idea that certain people and species belong in certain fixed places has had a long history in Western culture. Under its logic, migration is by necessity a catastrophe, because it violates the natural order.

That order had been defined hundreds of years earlier, by a sex-crazed Swedish taxonomist. Its foundational principle can be summed up simply.

We belong here.

*They belong there.*

# 3

# LINNAEUS'S LOATHSOME

# HARLOTRY

The son of an impoverished Lutheran minister and a rector's daughter, Carl Linnaeus was born in 1707 along the shores of a deep clear lake in southern Sweden, swaddled in a cradle decorated with blooms from his father's garden.

As a boy, Linnaeus spent his days walking in the lakeside woods, carefully examining the anatomy of plants and animals he found. Nature, for him, was a reflection of the Creator. And since the Creator was perfect, nature was perfect, too, with each living creature in its place with a specific function to fulfill. "Nature," according to Linnaeus, "never makes anything without a purpose." The beauty of it left him "completely stunned."

He grew up surrounded by human design and domesticated landscapes. The wild forests that had once dominated the region had long ago been razed and replaced with flat, arable meadows and orderly fields of grain. Around the rectory, Linnaeus's father created horticultural wonders. One of his gardens presented a botanical version of a fully laden dining-room table, in the form of a raised, circular garden bed with special plants and shrubs designed to represent the various dishes of a feast and its guests. Linnaeus spent hours playing in it. Later, his fans would call him the Prince of Flowers.

Order entranced Linnaeus, but as a natural historian he'd be called upon to describe the world's biodiversity in all its wild and dynamic chaos. Eighteenth-century society swirled with questions about the origins and distributions of living things on the planet and, with it, the history and nature of our similarities and differences and the role migration played in creating them.

Today such questions about the origin and distribution of species and peoples would be sequestered into a field known as "biogeography," a fascinating but mostly obscure branch of science generally considered of marginal public interest. Back then, biogeographical theory carried far-reaching consequences. The authority of the church; its hold on science, newly emerging from its shadow; the legitimacy of the colonial enterprise—and how generations of descendants would view and police migrants—all hung in the balance.

It would be up to natural historians like Linnaeus to provide answers to the most pressing questions of the day: Where did foreign peoples and strange species come from, and where did they belong?

∾

Bigger, faster ships with better navigation capacities had allowed European explorers to travel farther and longer than ever before, catapulting them deep into Asia, Africa, and the New World, where they encountered a previously unimaginable breadth of biodiversity. Companies such as the Dutch East India Company sent battalions of explorers and colonists into remote locales of the world to plunder resources, claim new territory, and establish new trade routes. Aspiring young naturalists joined them on years-long expeditions to the South Pacific and Asia.

They returned from their voyages overflowing with breathless tales about the bizarre-looking foreigners and creatures they'd glimpsed overseas. "Big fierce people, dark yellow in color," lived in the Nicobar Islands, recounted Nils Matsson Kioping, who'd visited the islands with the Dutch East India Company in the mid-seventeenth century. They wrung the necks of parrots and ate them raw, Kioping wrote. He's seen it himself when they swarmed and boarded his ship. Each "had a tail at the back, hanging like a cat's tail," he wrote. The celebrated writer François-Marie Arouet, who published under the name Voltaire, described a tribe of diminutive people with red eyes, who survived to only twenty-five years of age, living in Congo. "A very small and very rare nation," he explained. "Their minimal strength barely enables them to make their way out of the caverns where they live." These foreign peoples participated in strange, otherworldly practices. In parts of Africa, travelers revealed, whole tribes forced their males to undergo a ritual excision of one testicle, leaving them "monorchid."

Even when describing foreigners as recognizably human, eighteenth-century travel writers underlined the distinctions between European and non-European peoples and animals rather than their similarities. They described foreign peoples' skin colors not as a range of earth-toned hues but in crudely exaggerated categories of "red," "yellow," "black," and "white." They described the breasts of women in parts of Africa not just as "large" but as so ponderous that they had to be laid upon the ground first before the woman could lie down herself, and voluminous enough to be sold as tobacco pouches.

While presented as eyewitness accounts, these tales were mostly cobbled together from folklore, myth, and thirdhand gossip. Some of the most prolific authors, such as Arnoldus Montanus, who produced thousand-page illustrated tomes about the world outside Europe, had never even left the continent.

Voltaire's description of cave-dwelling peoples in Congo, for example, spun together various ancient myths. Herodotus had written of humanlike creatures he called "troglodytes" who lived in caves and fed on lizards. Pliny had contributed additional details about such beings, including that they were nocturnal, crawled around on their bellies like newborn puppies when exposed to sunlight, and made a "gnashing Noise" rather than speech. Voltaire cohered this mishmash into an actual people, attributing them to a location specific enough to seem credible—central Africa—and yet distant enough that few readers would ever be able to verify his claims for themselves.

Kioping's description of his encounter with the yellow tailed men might have similarly been based on myths. A three-foot-tall "hobbit" hominid, *Homo floresiensis* is now known to have inhabited islands around Nicobar as recently as thirteen thousand years ago, when humans lived there as well. It's possible that Kioping heard of such beings when he visited the region. Through generations of storytelling such a creature could have morphed into the tailed man he described, just as the Greek saint Nicholas had morphed into a North Pole–dwelling master of flying reindeer. He then retold the myth in the form of a dramatic personal encounter as a literary flourish.

Why were Europeans so struck by the differences they saw in their fellow peoples, rather than by the equally striking similarities they shared? It wasn't as if Europeans were some monolithically homogenous group of peoples, in contrast to peoples from other regions. Europeans themselves encompassed a wide range of hair types, skin tones, body shapes, and more. As a group, they were diverse, and they shared as many commonalities with peoples in other places as they did with one another. After all, peoples in Africa, Asia, and the Americas are and were kin.

One theory attributes Europeans' exaggerated perception of the strangeness of foreigners to the changing nature of travel at the time. Before the era of long-distance sea voyages, traders' and travelers' perceptions of differences between groups of people scattered across the landscape had remained indistinct. Europe's encounters with other groups of people—and theirs with Europe—had been the result of movements that were slow and plodding. Traveling by land, traders and explorers passed through contiguous, adjacent areas that shared overlapping geographical features and climates, as well as the usual conflict-ridden and romance-laden relations of human neighbors everywhere, whether enemies or allies. Bonded by shared climates and genetic relationships, whatever biological distinctions had arisen in one group gently graded into those that had emerged in the next. Those passing by would have seen a range of skin colors, body types, and facial characteristics grading subtly and perhaps imperceptibly, with few if any dramatic physical distinctions between groups.

Accordingly, in earlier eras, certain aspects of foreign people that appeared so strikingly distinct to eighteenth-century explorers— variations in skin color, for example—had been considered an irrelevant detail, like the pattern of spots on a dog. In the metropolises of Lower Nubia, Upper Nubia, and ancient Egypt, for example, people's skin tones ranged from fair to dark, in accordance with the fifteen degrees of latitude that the four-thousand-plus-mile Nile River valley they lived along crossed. But while contemporary artworks depicted their skin color diversity, skin color variation had nothing to do with the social hierarchy they'd maintained for thousands of years. In pre-Enlightenment Europe, too, artists and geographers tended to depict overseas peoples as physically similar to Europeans. A 1595 painting of the so-called Hottentots, a vaguely defined group of Africans, depicted them as two "classically Greek-looking men," as the biologist and historian Anne Fausto-Sterling points out. Skin color back then was more like hair color is today, a noticeable but socially meaningless detail.

The nature of eighteenth-century European exploration led to a distinctly different experience of human diversity. Instead of traveling through contiguous, connected regions on land, travelers journeyed thousands of miles over uninhabited seas. This had the effect of depositing them abruptly in entirely new regions with distinctly different climates and geographies. That may be why the continuity of human diversity may have appeared so strikingly discontinuous. It was as if, rather than wading from shallow warm waters into cooler deeper ones, they'd jumped directly into the depths.

Depictions of and stories about these strange foreign others—featured in illustrated volumes, paintings, tapestries, and other artifacts—dazzled, delighted, and confounded European sensibilities. Living specimens occasionally made their way into Europe in traveling exhibits, where even Europeans wary of arduous overseas travel could catch a glimpse of the oddities of the natural world beyond their shores. Wealthy elites created menageries in which they assembled live antelopes, lions, monkeys, flamingoes, and even more fantastical beasts. One exhibitor in Hamburg boasted of a seven-headed hydra in his collection, which natural historians traveled across the continent to view. It was a fake—an amalgam of body parts from weasels, glued together and covered in snakeskin—but the credulous public interest in the strange biology of foreigners was real. Exhibitors displayed women touted as mermaids, Hottentots, and troglodytes, who were often small African and South American children with albinism.

The point was not to accurately recount the details of foreign peoples and places. It was to express Europeans' ambient sense of foreignness. Their traveling exhibits of Hottentots and troglodytes and menageries full of seven-headed hydras were designed to shock—but they were also, in a way, an expression of shock itself. Whoever they were, whatever they looked like, the foreigners beyond Europe's borders were different: a breed apart.

This preoccupation with distinctions between peoples did not derive from any explicit consideration of migration. But to recognize the role of migration in our past, one had to accept the notion of our biological commonality. It is our shared humanity that makes our migratory past a logical necessity. How else could we have gotten around the world? The success of our past migrations, in turn, suggested the likelihood of similarly successful future ones. But with publishers and exhibitors lining their pockets with the most salacious and sensational depictions of foreign peoples, European perceptions of the oddity of foreigners steadily grew.

Debate swirled among intellectuals and elites gathered in newly formed scientific societies across the continent. Who had created specimens such as the yellow tailed men? The divine Creator, as the church said, or some other, unknown creative force in nature? Could such beings actually be related to Europeans, who commonly descended from Adam and Eve as the Bible indicated? And if so, how did they get to their far-flung locales where European explorers had encountered them? Most eighteenth-century explorers—despite having made the journey across oceans and continents themselves—could not imagine that anyone else might have done the same.

As the perceived gulf between peoples on different continents widened, the less credible the notion of a shared origin—and with it the promise of migration in our past and future—became.

～

Linnaeus had little direct knowledge of the extent of the world's biodiversity. He undertook his sole voyage of exploration while a medical student at Uppsala University. His itinerary was conservative: he would travel no farther than the northern province of Lapland, in his own Sweden. Nevertheless, he would still enjoy ample opportunities to learn about and understand the nature of cultural and biological diversity. The untamed northern tundra was poorly understood at the time, populated by reindeer-herding nomads whom Linnaeus called Laplanders, now known as the Sami. The Uppsala Science Society speculated that the Sami might be a lost tribe of Israel, or perhaps some mysteriously transplanted denizens of the New World. Some scholars hypothesized that they might be pygmies or the Central Asian nomads known as Scythians.

Linnaeus hired some guides and set off on foot for a six-month journey through Lapland. He took the safest route possible, clinging to the coast for as long as he could, scribbling notes on the flora and fauna and collecting unusual plants and insects. He was miserable. "How I wish that I had never undertaken my journey!" he wrote in his journal at one point. He "longed for a companion" and felt defeated by "this desolate wilderness." And he failed to learn much about the Sami. The "few natives I came upon spoke with a foreign accent," he complained.

Linnaeus wasn't much of an adventurer. He could not countenance people speaking languages other than Swedish. Later, when he'd be forced to visit Finland—he hadn't wanted to subject himself to the discomforts of travel—he privately complained that the people there didn't speak Swedish. "They speak nothing but Finnish," he noted with disdain. He also considered the Finns "quarrelsome" people and felt repelled by what he called a "disgusting stench of a sour white fish." One of his biographers, Lisbet Koerner, called him a "rude provincial—sentimental, superstitious, and devoid of general culture."

The trip was a failure. When he returned to Uppsala with great relief, he took pains to conceal its shortcomings. He submitted materials to his funders that exaggerated the hardships he'd encountered, even including details about one outing that required acts of physical prowess and daring so outlandish that modern biographers are certain that he made it up. He cobbled together an outfit including a Sami woman's cap and a drum, which he passed off as an authentic Sami costume and donned for special occasions. He even had his portrait painted while wearing it. He may not have learned much about the Sami, but nobody else knew much about them, either. For years he would present himself as such an experienced interlocutor of the Sami people that he'd practically become a Sami himself.

His benefactors were impressed. "I don't believe," one wrote, "there was a man so learn'd in all parts of natural history as he; and that not superficial, but to the bottom."

∾

In any discipline that attempts to create order from a confusion of data, there are what Charles Darwin would later call "lumpers" and "splitters." The splitters focus on differences between the data points, cleaving them into as many categories as necessary to distinguish each from the other based on their distinctions, however minute. Lumpers attempt to discern underlying similarities within the disparate data points, grouping as many together based on unifying commonalities.

Linnaeus, who would sniff out any hint of a distinction to draw yet another biological border, was a splitter.

Linnaeus started writing his groundbreaking taxonomy—a system of naming, describing, and classifying the world's biodiversity—while working as a personal physician and curator for a botanical garden estate owned by a director of the Dutch East India Company. He created a simple categorization system, one that anyone could use. He gave each species two Latin names: the first denoting its general category, and the second its specific character.

At first, Linnaeus left the thorny question of the origins and classification of foreign peoples unresolved. For many natural historians, the different shades of foreigners' skin—in particular the darker skin tones of African peoples—signaled some deeper physiological distinctions, the way the differently colored exterior of an apple distinguished it from, say, a pear. But Linnaeus struggled with how to fit that possibility into his taxonomy. If all peoples shared a common origin as the Bible said, then he'd have to admit that Europeans shared kinship with the foreigners they considered primitive and savage and possibly biologically alien. That was an unappealing option. At the same time, pointing to a separate lineage would suggest that the story of Adam and Eve was wrong, which was a sacrilege. Linnaeus sidestepped the issue. When it came to describing humans, Linnaeus wrote "*Nosce te ipsum*" by way of explanation: "know yourself." Basically, figure it out on your own.

Human bodies and relationships shaped his early taxonomies nevertheless. Recognizing the importance of sexual reproduction, he classified plants based on the anatomy of their sexual organs, categorizing male plants by their stamens and female plants by their pistils. Possibly because he could think of no other way of writing about it, he used the metaphors and language used to describe human sexual relations in his descriptions.

He described botanical marriages, husbands, wives, and harlots. He likened botanical sexual organs to those of humans. The anthers, pollen, and filament in stamens—the male sexual organ in plants—equated to the testes, semen, and vas deferens in human men; the style and tube of the pistil, pericarp, and seeds of plants' female sexual organs to the vulva, vagina, fallopian tubes, ovaries, and eggs in human females.

"Every animal feels the sexual urge," he wrote. "Yes, love comes even to the plants. Males and females, even the hermaphrodites, hold their nuptials . . . The actual petals of a flower . . . serving only as a bridal bed which the great Creator has so gloriously prepared, adorned with such precious bedcurtains and perfumed with so many sweet scents in order that the bridegroom and bride may therein celebrate their nuptials with the greater solemnity."

This steered Linnaeus into dangerous territory, for only a few plants conformed to the sexual practices that eighteenth-century Europeans considered respectable. Some female plants mated with twenty different male plants, and male plants mated with female plants other than their regular companions. Some plants reproduced with their own offspring. By inviting his readers to consider the reproductive act between a male and female plant as akin to a wedding night on an adorned bridal bed, Linnaeus implicitly invited them to consider these other much more provocative practices—incest, polygamy, adultery—in human terms as well.

The first edition of Linnaeus's *Systema Naturae* came out in 1735. Critics decried it as abhorrent, lewd, and vulgar. "Loathsome harlotry," the Prussian botanist Johann Siegesbeck roared. In one particularly cutting analysis that made the rounds, a critic used Linnaeus's sexual taxonomy to characterize Linnaeus himself as a plant-woman. "I was the laughing stock of everybody," Linnaeus complained. The opprobrium nearly drove him to a nervous breakdown.

~

Linnaeus's rival, the French naturalist George-Louis Leclerc, grew up on an estate in the village of Buffon, in the Dijon region of eastern France. The estate had been bought with a fortune his civil servant parents had inherited from his great-uncle. At university, he studied mathematics and medicine and traveled through Europe with his friend the Duke of Kingston. When he returned, he bought the village of Buffon, adding the suffix "de Buffon" to his name, and moved to Paris, where he'd been appointed as curator for the king's medical gardens.

If Linnaeus was a splitter, Buffon was a lumper. His ideas ravaged Linnaean taxonomy.

Unlike Linnaeus, who pictured nature as unchanging and rigidly ordered, Buffon saw it as mutable and dynamic. All of nature consisted of an unbroken continuum, separated only by "imperceptible nuances" and "unknown gradations," Buffon wrote. His vision of nature resurrected ancient ideas, such as those of the sixth century–B.C.E. Greek philosopher Heraclitus. The solidity of rock, the contours of waterways, and the habits of living creatures did not express some fundamentally unchanging material nature. They were just momentary expressions of processes in flux, with no fixed substance. Permanence was an illusion. What was real was change.

This led Buffon to some radical notions about human history and biology. All humans, regardless of where they lived or the color of their skin, Buffon wrote, "derive from the same stock and are of the same family."

If Europeans and Africans were biologically distinct like, say, horses and donkeys, the child of one European and one African parent would be sterile, like a mule. But such children weren't. "If the Mulatto were a real mule," he wrote, "there would indeed be two truly distinct species . . . and we would be right to think that the white and the nègre in no way had a common origin. But this presumption itself is refuted by reality."

What's more, Buffon knew of the phenomenon we now call albinism and that it occurred in dark-skinned Africans. Pliny the Elder, Ptolemy, and the Roman geographer Pomponius Mela had written about African people with albinism. The explorer Hernán Cortés had claimed to have encountered people with the condition in Montezuma's palace in 1519. This condition—in which dark-skinned parents produce pale-skinned progeny—was of great interest to eighteenth-century observers. Some commentators theorized that albinism was a kind of poxlike skin disease. Others argued that the African albino proved that the feature of pale skin had preceded that of dark skin. The albino African was like a wild offshoot of a culti-vated garden type, they argued, reverting back to ancestral type.

For Buffon, albinism among Africans proved that skin color was a superficial, mutable trait overlaid atop Europeans' and Afri-cans' common humanity. Buffon's friend Voltaire had written that "the Negro race is a species of men different to ours as the breed of spaniels is from that of greyhounds." But that couldn't be, Buffon pointed out. African parents, despite their dark skin, could give birth to pale-skinned babies. Spaniels don't give birth to greyhound puppies.

The observable differences between different peoples derived not from any intrinsic biological distinction, Buffon said, but from variable processes of change and adaptation.

~

Positing foreigners as human allowed Buffon to adhere to the biblical story tracing all of humanity back to the Garden of Eden.

But it did require him to explain how foreigners had disseminated themselves across the globe and, if they'd descended from Adam and Eve like the Europeans, how they'd acquired their dark skin and strange features.

He imagined a history of migration.

Buffon famously created labyrinths and mazes at the king's gardens. He envisioned a human past marked by similarly circuitous routes. There was no evidence of long-distance migrations at the time, but Russian expeditions had suggested a possible land bridge across the Bering Strait. Such a land bridge might have allowed people without the benefit of oceangoing ships to travel from the Old World into the New on foot, Buffon figured. Perhaps, he speculated, sometime in our deep past our ancestors had left the Garden of Eden on a series of long-distance migrations, depositing them in all the far-flung and diverse landscapes that European explorers had recently discovered.

Those migrations and dispersals—entirely theoretical though they were at the time—explained the distribution of humans in different parts of the world, as well as their various visual aspects that had so captivated and preoccupied eighteenth-century audiences. After migrating, Buffon speculated, peoples on different continents and regions adapted to a variety of unique environmental conditions, morphing their bodies into a range of shapes and colors.

The idea that weather patterns and climatic zones influenced health and the shape of the body dates back to Aristotle and Hippocrates. As scientists would later show, migration and the changes it forced does explain much of the observed variations that Europeans chattered about in their salons and scientific societies. The landscapes people migrated into left marks on our bodies. We evolved genes to help us digest the local foods, tolerate the local weather, and survive the local pathogens. To withstand

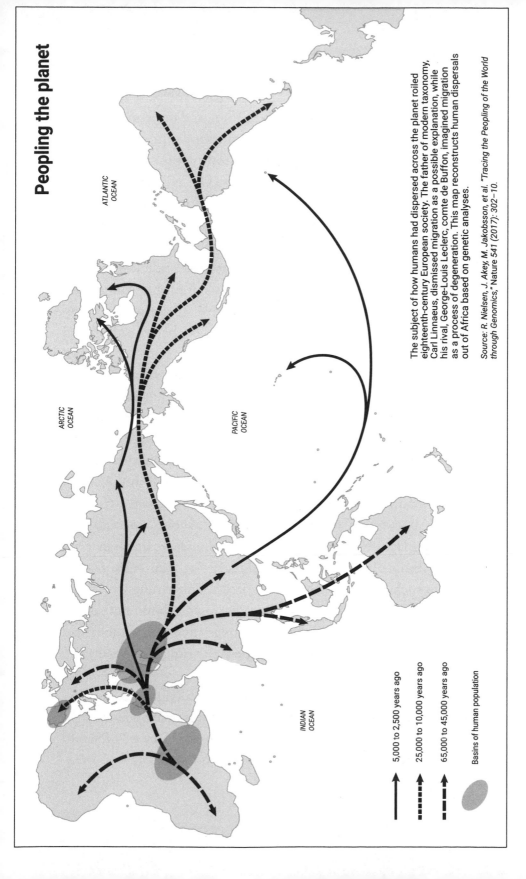

# Peopling the planet

ATLANTIC
OCEAN

ARCTIC
OCEAN

PACIFIC
OCEAN

INDIAN
OCEAN

The subject of how humans had dispersed across the planet roiled eighteenth-century European society. The father of modern taxonomy, Carl Linnaeus, dismissed migration as a possible explanation, while his rival, George-Louis Leclerc, comte de Buffon, imagined migration as a process of degeneration. This map reconstructs human dispersals out of Africa based on genetic analyses.

*Source: R. Nielsen, J. Akey, M. Jakobsson, et al. "Tracing the Peopling of the World through Genomics," Nature 541 (2017): 302–10.*

5,000 to 2,500 years ago

25,000 to 10,000 years ago

65,000 to 45,000 years ago

Basins of human population

the extreme cold of the Arctic, people had evolved higher meta-bolic rates, larger body mass, shorter limbs, and the stockier bodies that reduce heat loss. We evolved different skin tones. In the higher latitudes where vitamin D–bestowing sunlight is limited, we evolved paler skin that absorbed more UV light, and the ability to digest lactose, so we could ingest the vitamin D in milk. Those who migrated through the equatorial parts of the planet evolved the ability to retain the sodium that their bodies lost through sweat, and the long arms and legs that promoted heat loss and kept them cool. Many of the biological distinctions that eighteenth-century explorers noted, including variable skin colors and body types, did indeed stem, at least in part, from how bodies had adapted to different landscapes.

But Buffon is not remembered today for his prescience. Both he and Linnaeus drew on the conventional wisdom, passed down through philosophers and theologians since medieval times, that all matter and life on earth organized itself hierarchically, in what the ancients called the Great Chain of Being. Each kind or class of living thing or physical matter was its own link in the chain, its position representing its rank in terms of positive attributes. At the bottom of the chain lay rocks. Above emeralds and sapphires lay rubies and above them diamonds. A bit farther up, plants. Birds of prey occupied a higher rung than birds that ate worms such as robins, which occupied a higher rung than those lowly birds that ate seeds, such as sparrows. Above animals lay people. Peasants occupied the lowest rungs, then clergy, noblemen, and kings. Above them, on the uppermost rungs, lay angels and God himself.

And so for Buffon, differences between foreigners and Euro-peans that emerged via migration were not morally neutral.

As people and species migrated from the Garden of Eden, Buffon theorized, new diets and climates "degenerated" them.

Because the Garden of Eden had been located somewhere near Europe, Europeans retained much of their original state of perfection. He couldn't say as much for the Africans, Asians, and Americans, whose migrations into too-hot and too-cold climates had turned them into the misshapen, morally questionable creatures described by eighteenth-century voyagers.

The degenerative effects of migration explained the "savages" of the Americas, Buffon argued. They'd rotted in North America's wet and cold climate. "The Savage is weak and small in the organs of generation," he wrote. "He has neither beard nor attraction for the female . . . He is also less sensitive and more fearful . . . He will rest stupidly in repose on his haunches or sleep for the entire day . . . The most precious flames of the fire of Nature has been refused to them." He considered the colonists who'd settled in the Americas similarly compromised. Consider the lack of poets and geniuses among them, he pointed out. Even North American animals had become puny and diminished.

(The eleventh-century Islamic philosopher Avicenna had postulated a similar sequence of events, but in his scheme, it was the people in Europe who were mentally inferior to those in his own Central Asia. Deprived of sunlight, Europeans "lack keenness of understanding and clarity of intelligence," he'd held, while sun-drenched Nubians and Ethiopians "lack self-control and steadiness of mind." Both groups were best suited to enslavement, he figured.)

Buffon detailed his ideas in what would become a massive thirty-six-volume encyclopedia, *Histoire naturelle*, the first three volumes of which appeared in 1749.

Americans such as Thomas Jefferson disputed Buffon's unflattering portrayal of his countrymen and women as degenerates. Jefferson devoted a long chapter in his sole book, *Notes on the State of Virginia*, to debunking it. (His main counterevidence: the strapping bodies of moose, found only in North America.) But elsewhere, *Histoire naturelle* was a hit, read by every educated person on the continent.

Top scientific societies of Paris, Berlin, London, and elsewhere extended invitations for membership, and royals showered Buffon with gifts. The king made him comte de Buffon and commissioned a sculptor to create a bust of his likeness.

~

Linnaeus was not impressed.

"Wordy descriptions," he noted acidly. "Few observations . . . without any method," he went on. "Criticizes everyone, but forgets to criticize himself, although he himself has erred the most. Hater of all methods."

Over the years, Linnaeus had published additional versions of *Systema Naturae*, each one more elaborate than the last. Lewd or not, the system Linnaeus devised made the naming of living things uniform and universal. Before him, naturalists grouped aquatic mammals with fish, categorized four-footed animals according to their size, associated bats with birds of prey, and classified birds according to where they made their nests. They compared the specimens they found to illustrations in expensive, hand-colored and copper-engraved volumes to figure out if anyone had encountered them before, and if so, what they'd called them. But with Linnaeus's taxonomy, they didn't need that anymore. "Take a bird or a lizard or a flower from Patagonia or the South Seas, perhaps one that had had a local name for centuries, rechristen it with a Latin binomial, and presto!" writes the essayist Anne Fadiman. It became a tiny European colony. Linnaean taxonomy was a "form of mental colonising and empire-building," a potent tool in Europe's campaigns of conquest, writes the historian Richard Holmes. Any living creature anywhere could be fit into its order.

Luminaries and royal patrons from across the continent called on the celebrated naturalist, bearing gifts of exotic animals to add to his menagerie, which included cockatoos, peacocks, a cassowary, four kinds of parrots, an orangutan, monkeys, an agouti, and a raccoon. Students from all over Europe flocked to Uppsala University to hear Linnaeus hold forth on botany, natural history, diet, and disease. The luckiest among them might be subjects for one of his impromptu skull examinations, through which he determined the nature of their talents. Rather than quietly creep off into the forest to collect samples, as many field biologists might, Linnaeus, confident in his own celebrity, set off with his students in parades waving banners, playing horns, and beating drums, yelling "*Vivat Linnaeus!*"

Linnaeus had little truck for Buffon's reliance on migration as an explanation for the distribution of peoples and species. By elevating migratory transformation, Buffon's theories questioned the permanence of nature and challenged the perfection of the Creator.

Linnaeus discounted even the most obvious migrations in nature. To be fair, at the time, not much was known about wild migrations. Nobody back then could have easily tracked birds' movements over mountains and oceans, for example. Mariners had often reported seeing birds flapping across seas miles from shore, suggesting some secret wintertime passage. And some birds reappeared in the spring singing birdsong from Africa—the auditory equivalent of airline tags fluttering from their baggage. Some even turned up with bodies damaged by African-style spears, including one stork that arrived in Europe with an entire spear pierced through its body. (The Germans called her *der Pfeilstorch*, or "arrow stork.")

But possibilities other than migration could explain such phenomena. The theory that Linnaeus and many of his contemporaries subscribed to held that the reason birds disappeared in the fall was because they hid in caves, trees, and underwater for the winter. Aristotle had proposed that swallows hibernated at the bottom of lakes. This idea had been "treated as a matter of fact for hundreds of years," the historian Richard Armstrong explains. Linnaeus did not question it.

In the sixteenth century, the Swedish archbishop Olaus Magnus had illustrated his natural history tome, *History and Nature of the Northern Peoples*, with a depiction of a fisherman pulling a net full of sodden swallows out of the water, as if they'd been awakened from their submarine stupor. A seventeenth-century French ornithologist even went to the trouble of attempting to observe birds' winter hibernation, by watching over captured birds in his aviary all season to see if they fell into a seasonal slumber.

Regardless of the implausibility of winter hibernation, the alternative idea that birds annually traveled thousands of miles across continents and oceans clashed with the Christian paradigm of an unchanging, orderly world. Heraclitan ideas had been condemned as pagan and backward by the church: the third-century Christian theologian Hippolytus called Heraclitus's notions "blasphemous folly." If creatures migrated great distances between continents, where had the Creator "fixed" them and into which place? What could possibly be the purpose of departures for distant foreign climes when the surrounding natural world was so stable and harmonious? In the Bible, after all, migratory creatures expressed God's divine punishment, not his perfection. Take migratory insects, for example, which are the most cited insects in the Bible. God sends them out as plagues to punish the early Egyptians, as a curse for disobedience, and as harbingers of apocalypse.

For Linnaeus, there'd been only a single dispersal in the past. He imagined the Garden of Eden as a tropical, mountainous island in a primordial sea, where cold-weather creatures lived atop the summits while warm-weather ones confined themselves to the plains. As the sea receded, the original animals and plants slowly dispersed from Eden into their current locations in the cold and warm parts of the earth. The dispersal happened once at the beginning of time. Since then, no species ever arose nor was extinguished. "It is impossible," Linnaeus wrote, "that anything which has ever been established by the all-wise Creator can ever disappear." Nor did they ever change. That was a logical corollary to the Creator's perfection and all-knowing omnipotence.

He dismissed Buffon's work, and his ideas about migration, change, and climatic adaptations, out of hand. "As he is rather eloquent that seems to count for something," Linnaeus grumbled, but Buffon "isn't particularly learned."

Buffon's method of cataloging nature with a lush descriptive method that emphasized dynamism and fluidity was shallow and pretentious. Even worse, Linnaeus noted, Buffon had written his encyclopedia "in French," which Linnaeus frowned on and could not read. For Linnaeus, everything about Buffon's theory and the way he described it stank of swanky Parisian elitism.

He named a plant after the comte, a foul-smelling weed he dubbed "Buffonia."

~

With his tenth and most authoritative edition of *Systema Naturae*, Linnaeus crushed Buffon and his ideas for good. In it, he named and classified over four thousand animals and nearly eight thousand plants. He also laid out a definitive human taxonomy, settling the question of the differences between foreigners and Europeans once and for all.

While foreigners were popularly understood to be, in some inchoate way, biologically distinct from Europeans, evidence that any observed differences extended beyond the most superficial was spotty at best. For nearly a century, European microscopists and anatomists had searched for systematic biological explanations for the observed physical distinctions between Europeans and other peoples. Despite decades of research, the best evidence dated back to 1665, when the microscopist Marcello Malpighi had claimed to discover, between Africans' darkened outer layer of skin and the inner white one, a third layer of skin, which Malpighi called the "reticulum mucosum" or "Malpighian layer." According to Malpighi, this novel physiological feature, found exclusively in African bodies, consisted of a thick, fatty black liquid of unknown provenance.

The Malpighian layer was taken up as the smoking gun that proved that Africans were, in fact, biologically distinct from Europeans. "The mucous membrane, or network, which Nature has spread between the muscles and the skin, is white in us and black or copper-colored in them," Voltaire wrote. But upon further investigation, the Malpighian layer was revealed as phantasmal. It was impossible to extract the thick, fatty black liquid itself, as the French anatomist Alexis Littré discovered in 1702, when he attempted to isolate the layer's gelatinous substance by soaking African skin in various liquids. (He had also searched for the source of blackness in the sexual organs of an African man, which he dissected.)

In 1739 a French scientific society called the Académie royale des sciences de Bordeaux had laid down a challenge for the scientific community: "What is the physical cause of the *nègres'* color, or the quality of their hair, and of the degeneration of the one and of the other?" the Académie demanded, offering a prize to whichever natural historian could come up with the best answer to the question.

The Dutch anatomist Antonie van Leeuwenhoek's inquiries led him to believe that the color of African skin derived from darkened scales. Or perhaps, as the physician Pierre Barrère had surmised through his dissection of African slaves, it radiated from darkened bile inside the body, which stained both its tissues and the skin. None of it amounted to anything definitive. One Parisian anatomist examined the skin of an African man, using a chemical compound to blister and remove it. He found a dark-colored outer layer, unsurprisingly, over a white inner layer. What did it mean about the extent and origins of the most noted physical difference between Europeans and foreign peoples from Africa? Not much. Even Europeans came in a variety of skin tones. He surmised that the sun had seared their skins.

The Académie question remained unanswered, but in the end that wouldn't matter to Linnaeus. A different biological feature, albeit equally elusive, would prove far more influential in his assessment of which peoples belonged where.

~

Sexual anatomy fascinated Linnaeus. Variations in reproductive organs formed the basis of his taxonomic system. But not only that. He'd elevated the breast as the common feature distinctive to the category of creatures he named after the Latin *mamma* for "breast" as "mammals," rather than their other shared features, such as their distinctive hair, jawbones, or additional characteristics. He'd penned a special book for his son that included clinical details on adultery, incest, masturbation, and such varied topics as how women can make intercourse unpleasant for their male partners. When animals in his menagerie died, he routinely dissected their genitals, as one of his biographers notes.

Any hint of a difference in reproductive organs in peoples would figure prominently in his taxonomic scheme. And according to contemporary reports, foreign bodies—in particular those of women from parts of Africa—did indeed have unique reproductive organs, including a body part that no European possessed.

It was known as the "Hottentot apron," the "sinus pudoris," or the "genital flap." It was first reported by the French explorer François Le Vaillant. Translators deleted Le Vaillant's description of the feature in the English translation of his writings, but his illustration, supposedly based on his own observations, was widely circulated. It depicted a naked woman with two long, skinny tails, reaching about to her knees, hanging from her labia minora.

At first, European travelers speculated that the sinus pudoris was an artifact of genital mutilation. Le Vaillant claimed it was a form a fashion; the Dutch East India Company seaman Nicolaus de Graaf described it as a bodily "ornament." But as the eighteenth century progressed, and speculation about biologically distinct foreigners intensified, European naturalists increasingly understood the sinus pudoris as an authentic body part. Buffon described it as an "outgrowth of wide and stiff skin that grows over the pubic bone." The French zoologist Georges Cuvier considered it to be evidence of the nonhumanness of Africans. The Hottentots' genitals were similar to those of "female Mandrilles, Baboons, etc., . . . which take at certain times of their life a really monstrous increase," he wrote.

The sinus pudoris, along with other anatomical differences between Europeans and foreigners, formed the borderlines defining Linnaeus's human taxonomy.

Certain humans, he said, were a separate species altogether. *Homo troglodytes*, he wrote, is "certainly not of the same species as man, nor of common descent or blood with us." *Homo caudatus*, he wrote, is an "inhabitant of the Antarctic globe" that "can strike fire, and also eat flesh, although it devours it raw." *Homo caudatus* included the tailed men of Borneo and Nicobar, whom he'd read about in Kioping. The Sami, peoples he'd spent months with, he categorized as a nonhuman species, too: *Homo monstrosus*, a group that included dwarfs and Patagonian giants. (Buffon categorized the Sami, under his theory of degeneration, as "dwarfish degenerates.")

The human species, too, fell into distinct biological categories—subspecies even—each homogenous and specific to its own place in the landscape and in the moral order.

*Homo sapiens europaeus*, the peoples of Europe, were "white, serious, strong," with flowing blond hair and blue eyes. They were "active, very smart, inventive," Linnaeus wrote in his taxonomy. "Covered by tight clothing. Ruled by laws."

The people who lived in Asia were a separate subspecies called *Homo sapiens asiaticus*. "Yellow, melancholy, greedy," he wrote. "Hair black. Eyes dark. Severe, haughty, desirous. Covered by loose garments. Ruled by opinion."

The peoples of the Americas were a subspecies called *Homo sapiens americanus*. "Red, ill-tempered, subjugated," Linnaeus wrote in his description. "Hair black, straight, thick. Nostrils wide; face harsh, beard scanty. Obstinate, contented, free. Paints himself with red lines. Ruled by custom."

And finally the most distinct subspecies of all was *Homo sapiens afer*, the peoples of Africa. Linnaeus speculated, privately, that this subspecies might not be fully human but descended from a cross between a human and troglodyte. "Black, impassive, lazy," his taxonomy read. "Hair kinked. Skin silky. Nose flat. Lips thick. Women with genital flap; breasts large. Crafty, slow, foolish. Anoints himself with grease. Ruled by caprice."

With this human taxonomy, Linnaeus proclaimed natural history's independence from church teachings, disentangling science from religion and allying it with the state, a realignment that would make possible the rise of modern scientific authority. While possibly sacrilegious, his idea that humans fell into biologically distinct groups fixed to separate continents, each in its place—blacks in Africa, reds in America, yellows in Asia, and whites in Europe— facilitated Europe's political and economic interests. If foreigners were kin to Europeans, as Buffon had argued, then an argument could be made that they deserved the same rights, privileges, and moral consideration as anyone else, an argument that would pose a serious impediment to colonial designs on foreign lands and bodies. From a colonial perspective, it was more convenient to cast foreigners as so strange as to be unrelated or perhaps not even human at all. When the Dutch first settled southern Africa, for example, they'd considered the local peoples whose lands they invaded not as humans but as animals. They even claimed to shoot and eat them on occasion. Now such activities had the imprimatur of the world's most famous natural historian.

The publication of his tenth edition of *Systema Naturae*, including his definitive human taxonomy, ushered in Linnaeus's "rapid historical triumph" over Buffon, the science historian Phillip R. Sloan writes. Influential eighteenth-century writers rejected Buffon's theory of degeneration, adopting instead Linnaeus's concept of foreigners as different subspecies of humans, color-coded by continent.

In 1774 Louis XV ordered Linnaeus's classification system to be officially adopted. Jean-Jacques Rousseau claimed he knew "no greater man on earth"; Goethe, that only Shakespeare and Spinoza had been more influential on his thinking. In 1776 the Prince of Flowers was ennobled as Carl von Linné.

In Linnaean taxonomy, nature existed in discrete units, defined by biological borders. Each creature and people survived in its own place, separate and isolated from the others. The connective tissue that migrants created between peoples and places played little biological role of note. It barely existed.

As Linnaeus ascended to his place in history as the Father of Modern Taxonomy, migrants and migration as a force in nature and history receded into the background.

~

The most explosive claim in Linnaean taxonomy, that people who lived on different continents were biologically foreign to one another, a claim that would fuel centuries of xenophobia and generations of racial violence, rested on a single body part, the sinus pudoris. But very few—quite possibly none—of those who commented on the sinus pudoris had actually ever seen it.

Linnaeus hadn't. He had tried to catch a glimpse of one on his visit to Lapland. When an elderly Sami woman in Lapland sat casually in front of him wearing a short dress, he took the opportunity to jot down a "detailed description of her pudendum" or vulva, one of his biographers noted. (This particular stretch of Linnaean insight has yet to be translated from the Latin.) He'd written to the Swedish East India Company, asking them to acquire a "troglodyte" for him so he could personally examine it. (He'd similarly pleaded with the Swedish Academy of Sciences to sell him—alive or preserved—a Danish mermaid that they claimed to have on exhibit in Jutland. "This is a phenomenon which does not occur more than once every 100 or 1,000 years," he explained.)

A traveling exhibit of a troglodyte arrived in London right around the time he wrote his human taxonomy. He had "never been so delighted," the science historian Gunnar Broberg writes. Linnaeus heard that the creature—in fact, a ten-year-old girl from Jamaica with albinism—was "wholly white, but with negroid features," and in addition had "pale yellow eyes turned to a curious position as if squinting, and unable to tolerate daylight, although seeing better in the dark."

First he tried to buy the girl and bring her to Uppsala. When that failed, he dispatched one of his students to London with instructions to closely examine her genitals. But despite Linnaeus's promise of a membership in the prestigious Society of Science at Uppsala if he succeeded, the student came back empty-handed. The young girl's keeper refused to cooperate with his request, regardless of the stature of the famed naturalist who'd sent him.

European scientists continued to be foiled for decades. In 1810 a Dutch businessman brought a woman named Saartjie Baartman, who'd worked as a servant for Dutch farmers near Cape Town in South Africa, to Europe to put on exhibit. He called her the Hottentot Venus. Her tour through the capitals of Europe attracted widespread interest. During the exhibit, Baartman would be "produced like a wild beast, and ordered to move backwards and forward and come out and go into her cage, more like a bear on a chain than a human being," as antislavery activists put it at the time. Viewers could, for an extra fee, poke and prod her behind.

Europe's most famous scientists flocked to view the exhibit in hopes of confirming the fact of her sinus pudoris. Charles Darwin's cousin, Francis Galton, visited the exhibit equipped with his sextant, which he used to measure her body from every direction so he could figure out her precise dimensions. But while she was on display in the traveling exhibit, a fig leaf generally covered the organ in question.

Cuvier arranged for a commission of zoologists and physiologists to examine Baartman during a three-day scientific survey in Paris. During her examination, Baartman clutched a handkerchief over herself, only briefly and with "great sorrow," as the science historian Londa Schiebinger puts it, allowing it to drop. The gathered scientists didn't see anything out of the ordinary, but they figured she hadn't given them enough time to get a good look.

Baartman died in 1815, at the age of twenty-six. Cuvier seized on the opportunity to dissect her body and capture the holy grail of eighteenth-century anatomy once and for all.

He didn't find anything like what Le Vaillant and Linnaeus had described. There was no long skinny tail attached to her genitals. All he found were pretty run-of-the-mill labia. They looked to Cuvier like "two wrinkled fleshy petals." The most he could say about them was that they seemed "greatly enlarged."

Still, absence of evidence was not taken for evidence of absence. Instead, Cuvier presented his result with all the fanfare of a most significant and telling finding. He devoted nine pages of his sixteen-page memoir about the dissection to a detailed description of Baartman's genitals, along with her breasts, buttocks, and pelvis, paying homage to Linnaeus's presumption of her divergent sexual anatomy. He removed Baartman's genitals from her body and preserved them in a jar at the anthropology museum Musée de l'homme in Paris.

For decades, museums and exhibitors displayed Baartman's body as proof of Linnaeus's characterization of non-Europeans as biologically alien. Plaster casts of her body, enlarged illustrations, and even a stuffed display of her actual skin appeared in museums and exhibits across the continent, including at the 1937 Paris International Exposition, where tens of millions marveled at her ordinary human body, as if it were somehow distinct from their own.

The paleontologist Stephen Jay Gould wrote about his visit to the Musée de l'homme in 1987. He found the jar of preserved genitalia Cuvier had prepared sitting on a shelf in the basement.

∾

Linnaean taxonomy formed the basis for the modern study of nature. Later taxonomists updated his classifications but maintained its basic structure. Linnaeus's system of reflecting the geographic location of a species in its name "became unreliable," his biographer Lisbet Koerner writes, as scientists discovered that most were more "divergent and geographically dispersed" than Linnaeus had presumed. They moved around. But his categorization of insects into flies, bees and wasps, butterflies, lacewings, bugs and aphids, and beetles, held up for years even as scientists discovered hundreds of thousands of new insect species. His human taxonomy proved equally influential, although less heralded. Linnaeus was not bold enough to make the heretical argument that the various human subspecies could not possibly have commonly descended from Adam and Eve. Crossing the church risked royal censorship. The eighteenth-century naturalist Pierre-Louis Moreau of Maupertuis, for example, who had similarly described Africans as a separate species from Europeans, responded to the question as to whether such strange foreigners could descend from the same mother with "*Il ne nous est pas permis d'en douter*": "We are not allowed to doubt it."

Linnaeus did not dare doubt it either, but at the same time, he didn't bother reconciling his depiction of foreign human subspecies with the Bible. He issued his human classification system and let others interpret it.

While Linnaeus refused to spell out its implications, bolder scientists would. Migration had played no role in disseminating peoples around the planet, they said. There was no common ancestry between peoples. Foreigners were biologically alien, as different from native peoples as cats from dogs.

During Linnaeus's time, these notions did not impinge on most people's daily lives. Most people did not freely mix with people who'd been born on different continents. That would change when transatlantic shipping brought masses of people from Europe, Asia, and Africa together in the New World. People from distant places would not just glimpse each other from afar or read about each other in stories. They'd brush against each other in alleys, drink at the same bars, and work alongside each other on factory floors. They'd fall in love. They'd have babies.

Scientists predicted a biological disaster, igniting a social panic that would shape scientific inquiry, law, and politics for decades.

# 4

# THE DEADLY HYBRID

On the streets of early twentieth-century New York, the bodies of foreigners and natives collided daily, whether they wanted to or not.

Over a century and a half had passed since Linnaeus failed to entice a foreigner to draw close enough for him to touch. Since then Europeans had captured and shipped over 12 million people from Africa into the Americas to serve as their slaves, treating them as the subhuman entities that Linnaeus had described. African Americans had started trickling out of the slave-owning cities and towns of the U.S. South nearly as soon as their forced migration across the Atlantic ended. But after slavery was abolished, that trickle grew into a stream and then a river.

Over five hundred thousand African Americans fled the South in the first decade of the twentieth century. During the 1920s, over nine hundred thousand blacks migrated out of the South; in the 1930s, nearly five hundred thousand did. Ultimately over 6 million would flee the South. Their migration transformed the country. Chicago, which began the twentieth century with a population that was less than 2 percent black, would become one-third black by 1970; Detroit's black population swelled from 1.4 percent to 44 percent.

At the same time, people from Europe, Asia, the Caribbean, Central America, and elsewhere streamed into the country to seek cheap farmland or factory jobs, to pan for gold, and to escape bloody revolutions. Between 1880 and 1930, over 27 million entered the United States. Every week steamships pulled into New York City's ports to disgorge tens of thousands fleeing famine, poverty, and persecution in Ireland, Poland, Russia, and elsewhere. Over the course of the 1870s, about 3 million migrants arrived in the city; over just three years in the 1880s, another 3 million came. In 1890 a special station had had to be built, at Ellis Island, just to process them all.

The newcomers took jobs peddling used clothes or shad and clams, or they toiled as shoemakers and dockworkers, sewing their bedraggled children into their winter clothes for the season, and retiring to the city's windowless tenements and immigrant board-inghouses by night. In the city's filthy, polyglot streets and dance halls, African Americans newly arrived from the South and immigrants from across the Atlantic rubbed shoulders, creating a new mongrel culture replete with its own dance forms, such as tap dancing, a combination of Irish jig dancing and African American shuffling.

The lifestyle of the old New York elites who lived in stately homes in southern Manhattan and picnicked on Bunker Hill in the summer vanished. Developers tore down grand old houses to make way for tenement buildings. The cultural and demographic dominance of families like those of Henry Fairfield Osborn and Madison Grant, whose ancestors had settled the city when it was just a sleepy port town, eroded. By the turn of the twentieth century, immigrants and their progeny outnumbered people whose parents had both been born in the country. Of the 1.8 million people who lived in New York City, 1.4 million had at least one parent born outside the country.

Osborn and Grant belonged to an elite circle of educated, aristocratic New Yorkers. The son of a railroad tycoon, broad-shouldered Osborn kept a neat mustache and had deep-set, penetrating eyes. He'd trained at Princeton as a geologist and paleontologist and punctuated his life in the city with elaborate trips to remote locales. One of his paleontological expeditions had famously led to his naming and description of *Tyrannosaurus rex* and *Velociraptor*. His friend Grant traced his aristocratic heritage back to seventeenth-century Huguenots and Puritan settlers and favored big-game hunts with buddies such as Theodore Roosevelt. As a wildlife enthusiast, Grant would eventually help establish Glacier and Denali National Parks. He'd even have a species of caribou, *Rangifer tarandus granti*, named after him.

The transformation of the city undoubtedly rankled Grant and Osborn on a number of levels. But the two friends prided themselves on being "scientific men," people with either a stake or credentials in the increasingly prestigious and male-dominated world of scientific inquiry. It was the biological implications of immigration that would shock them into action.

$\sim$

Grant and Osborn wielded outsized influence over how early twentieth-century Americans understood biological science. Besides being "scientific men," they were science popularizers. Grant helped found the Bronx Zoo and belonged to a number of influential scientific and conservation societies. Osborn presided over the American Museum of Natural History. He was world-renowned for his displays—murals, dioramas, and mounted skeletons—that lured millions into the museum's cavernous exhibit halls.

Like other scientific men of the early twentieth century, both Grant and Osborn recognized the biological challenge posed by people of African, Irish, Polish, Russian, and Italian descent who crowded into New York City's tenements and slums.

Over the course of the nineteenth century, leading scientists had upgraded Linnaeus's theory of human subspecies, though based on a mix of secondhand gossip, folklore, and fabricated body parts, into scientific truth. In 1850 one of the era's most influential scientists, the Harvard University zoologist Louis Agassiz—he founded Harvard's Museum of Comparative Zoology in Cambridge, where streets and schools have been named after him—proclaimed as much. "Viewed zoologically," he'd told fellow members of the American Association for the Advancement of Science, "the several races of men . . . are well marked and distinct." Agassiz and other scientists had disseminated Linnaeus's myth of human subspecies in textbooks such as the 1853 best seller, *Types of Mankind*, and in photographic collections that depicted the various human subspecies much as pictorial charts depicted different animal species. Agassiz himself had commissioned several collections of images of disrobed bodies of enslaved Africans in South Carolina and workers in Brazil, which he presented as visual archives of the world's "pure racial types."

Naturalists had become so convinced of the reality of human subspecies that a whole new field of inquiry—"racial science"—had sprung up to refer to their biology. Just as herpetologists detailed the biology of reptiles and entomologists that of insects, race scientists detailed the biology of human subspecies or races. Aware that skin color could be subjective as a biological borderline between the races, race scientists searched for other biological markers that could be used to distinguish human subspecies from one another, just as different patterns on butterfly wings could be used to distinguish a monarch butterfly from a viceroy. They said that each subspecies had a distinctive "cephalic index," that is, the ratio of a skull's maximum length to its maximum breadth, multiplied by one hundred. They said that each had a specific "sitting height index," which could be calculated by dividing the median sitting height by median stature. According to their data, the measure averaged to 50.5 in Africans, and 53.0 in Americans. In 1900, when scientists discovered that human blood consisted of different varieties of blood cell types, they speculated that these distinctions would be found to be specific to human subspecies.

The political and economic value of their research was clear: scientific proof of a racial hierarchy justified the race-segregated economy at home and colonial conquests overseas. But race scientists struggled with an onslaught of messy contradictions in their data. As later scientists would confirm, thanks to our common ancestry and our border-crossing tendencies, the differences between human populations are superficial and fleeting. Through trade, capture, and conquest, people from different cultures and continents continuously collided, melding cultures and sharing genes, blurring the distinctions between us. Race scientists hunted for borders that were fuzzy at best. Even the cephalic index, which they considered the most authoritative measure of the human subspecies distinction, failed on a number of fronts. People from Turkey, England, and Hawaii, for example, often had identical cephalic indices, although according to race science they hailed from different races. People from isolated populations didn't have more homogenous cephalic indices than those from more mixed populations, although according to race science, they should have.

These anomalous results simply hardened race scientists' resolve to gather yet more data and devise yet more standards to pin down the biological border between peoples that they felt certain existed. They didn't force a course correction. Neither did the counterarguments of the scientist who would ultimately revolutionize biology.

Charles Darwin had purposely omitted any mention of human evolution in his *Origin of Species*, which appeared in 1859. Like Linnaeus, he felt trepidatious about spelling out how his ideas reflected on human society for fear of the political firestorms that might erupt. His theory of evolution had not found a particularly receptive audience. The year after his paper on evolution was read at the Linnean Society of London, the president of the society said that the previous year had included no revolutionary discoveries. Agassiz dubbed the book a "scientific mistake, untrue in its facts, unscientific in its methods, and mischievous in its tendency."

For Darwin, differences between peoples were nothing like the differences between zoological species, as Agassiz had said. Any child could tell the difference between a dog and a cat, he pointed out, but they would have to be instructed in order to perceive the minute differences attributed to race. If human subspecies distinctions were as biologically significant as Agassiz and others claimed, Darwin said, they'd be more like the tail of a tiger or the patterns on a butterfly wing: biologically fixed in each subspecies. They weren't. Finally, true subspecies don't inadvertently fuse when sharing the same territory, something that happened with human "subspecies" all the time, as especially evident in Brazil, Chile, Polynesia, and elsewhere. Like Buffon, Darwin felt that the minor differences between peoples derived from easily mutable adaptations to local conditions, like diet and climate. Such differences, he thought, could become exaggerated by local sexual preferences.

But as the race scientists grew more confident, Darwin grew less so. Writing became a struggle. He suffered "hysterical crying" and "dying sensations," as his biographers put it. The longer he delayed publishing his ideas on human diversity, the worse it got:

race science grew more powerful and his own ideas about a single human family more subversive. Worse, Darwin's efforts to acquire data from the East India Company and from various army surgeons in the British colonies, which would undermine the human subspecies theory, failed.

By the time he published *The Descent of Man* in 1871, presenting his arguments against the concept of human subspecies, over a decade had passed since the release of *Origin of Species*. It was, by then, too late. Darwin's influence on the scientific establishment had fallen into seemingly terminal decline. Leading nineteenth-century scientists had judged his ideas marginal and irrelevant. Darwin was an "ignoramus," the German physician Rudolf Virchow said. "The man is clearly crazy," Josiah Clark Nott, prominent yellow fever researcher and founder of the University of Alabama medical school, added. One attack on his ideas had been titled "At the Deathbed of Darwinism."

Darwin's *Origin of Species* would be resurrected, decades later. But the famous biologist's views on the nonexistence of human subspecies faded into obscurity. His biographers would call *Descent of Man* "Darwin's greatest unread book."

And so science popularizers such as Osborn and Grant justifiably showcased the findings of race science as established fact, not contested theory. At the Museum of Natural History, curators set up a "Hall of the Age of Man," such that visitors could physically walk the path of evolutionary progress. It ended with a display on the biological divisions in humans, and the hierarchical evolutionary relationships between different peoples, called "Races of Man." At the Bronx Zoo, Grant assembled more visceral exhibits based on the insights of race science, such as Linnaeus's characterization of *Homo sapiens afer* as only partly human. In one, his curators caged a man from Congo named Ota Benga in the monkey house. From the other side of the bars, zoo visitors could watch

Benga cavort with an orangutan and examine, confusedly, a pair of canvas shoes. With every chortle from a bemused visitor, our shared history as a single migratory species, and the superficiality of our differences, sank below the horizon.

～

Besides disseminating the insights of race science, Grant and Osborn also worked to promote new ideas about biological inheritance. Social reformers at the time advocated for improvements in sanitation, nutrition, education, and health care, which they said would uplift the strength and intelligence of the population. The latest findings about biological inheritance, Grant and Osborn felt, suggested otherwise. They also deepened scientific concern about the biological perils of migration.

Expert opinion about inherited traits remained unsettled through most of the nineteenth century. The so-called blending hypothesis posited that the qualities of each parent "blended" in the offspring, like chocolate milk swirling into plain. That certainly happened, but at the same time, it couldn't be the whole picture. When traits blend, the tall mother and diminutive father produce a brood of medium-height children. But if blending were the sole process in inheritance, after a sufficient number of generations, there'd be no short or tall people left at all, which was clearly not the case. Others believed that the qualities passed on from one generation to the next could be altered during an individual's lifetime. Buffon's protégé, Jean-Baptiste Lamarck, posited that giraffes could evolve long necks simply by spending much of their time stretching to reach the leaves of the treetops.

In 1899 the embryologist August Weismann refuted both theories by methodically removing the tails of five generations of white mice.

If the qualities of the parents are blended in the offspring, as the blending hypothesis suggested, or if environmental conditions had any effect on the traits they passed from one generation to the next, as Lamarck and others argued, then he'd see some inherited effect of the ritualistic tail-chopping. Over the course of several generations, the tailless mice's offspring would be born with no tails, say, or at least with shortened ones. But they hadn't been. Each subsequent generation developed normal tails, with no blending or environmental effect at all.

Not long afterward a few botanists in Europe published papers resurrecting some obscure experiments conducted decades earlier by an Augustinian monk named Gregor Mendel. Mendel had conducted tens of thousands of experiments in pea plants, carefully recording how traits such as whether peas were wrinkled or smooth traveled through the generations. He, too, had found that rather than blending with other traits or varying according to environmental conditions, traits marched unchanged from generation to generation, expressing themselves based on a single, intrinsic, and immutable factor: whether the trait was "dominant" or "recessive."

Mendel's work appeared to validate the rigid process suggested by Weismann's results. A new theory was born, "Weismannism," according to which inherited traits advanced through the generations like stones passing through a gullet, impervious to external conditions or the influence of other traits.

Weismann's experiments did not, by themselves, prove anything about the complex ways inherited traits changed as they passed through the generations nor about the effects of environment on the process. In fact, inherited traits and the genes that shaped them mix, match, recombine, and reassort in all kinds of multifarious ways, and a wide range of environmental effects influence the way they express themselves in our bodies, as geneticists would later learn. And Mendel's experiments, while shedding light on one form of inheritance, was only a tiny part of the overall picture. Genes did all sorts of different things and expressed themselves in all sorts of different ways besides the simple mechanism he'd discovered.

Nevertheless, scientists were able to collect data supporting the idea that Weismannism functioned in people, too.

In humans only a few traits follow the Mendelian pattern, such as eye color; an enzyme deficiency called alkaptonuria, which blackens the color of urine; and to some extent, hair and skin color. That's not to say that complex traits such as academic achievement or athletic prowess or economic wealth are not passed down from generation to generation. They are, but through cultural and economic processes, not biological ones. Because scientists did not distinguish between traits passed down socially and those passed down biologically, they claimed they could detect a Weismannist process in a range of complex traits as well. They had a simple method: scientists would pick a trait, figure out who had it, and then track its progress through the generations, either in real time or using genealogical records.

Darwin's cousin Francis Galton, for example, studied one thousand "eminent" men and their relatives, finding that the trait of "eminence" passed down through the generations, exactly as the trait of wrinkledness had passed through Mendel's pea plants. The zoologist Charles Davenport, who wrote an influential textbook on the topic, claimed through his studies of genealogy that traits of "quickness and activity in movement," "fluency in conversation," and the ability to learn new languages clustered in certain families, as did traits such as being able to "whistle a tune or sing a song without any apparent effort," which he took as proof that these, too, passed down through the generations biologically.

Weismannism electrified the scientific community. The old ideas, while incomplete, had properly cast inheritance as a mysterious, mutable process, almost impossible to fully control. Weismannism suggested that scientists could not only decipher the inheritance process but master it and thereby shape the fate of the nation.

Weismannism meant that intelligence, moral strength, musicality, and other socially beneficial qualities did not have to be carefully nurtured with good nutrition or enlightened education or moral instruction, as the social reformers said. It simply had to be endowed as a biological gift to future generations, like a strong nose or a weak chin. So long as those with the best traits had the most children, society would be assured of a brilliant, beautiful, morally upstanding populace.

Galton spearheaded a new movement to urge policy makers to reorient programs of social betterment based on the new science of inheritance. He called it "eugenics," for *eu*, or "good," with *genesis*. Instead of devoting resources to improving schools and nutrition, eugenicists said, policy makers should instead focus on who had sex with whom. Osborn and Grant agreed, founding the Galton Society to spread the eugenic gospel in the United States.

At the time, nobody knew what the mysterious matter that traveled through the generations consisted of. It would be years before scientists fingered DNA as the source of biological inheritance and began to comprehend the multifarious ways it functioned in the body and in relation to the environment. People like Osborn and Grant knew only that an enigmatic material, which they called, variously, "Mendelian factors" or "germplasm," endured. Osborn called it "the most stable form of matter which has thus far been discovered."

~

Twice a year Osborn and Grant donned their tuxedoes and white ties and headed to gatherings of the exclusive Half-Moon Club, where they quaffed gin with fellow members and listened to guest speakers talk about their latest conquests in the world of scientific exploration.

At one such gathering, they heard the Massachusetts Institute of Technology economist and race theorist William Z. Ripley deliver a lecture called "The Migration of Races."

In it, he spelled out the implications of Weismannism and race science on societies experiencing mass immigration from distant continents such as their own. It wasn't just that the newcomers would overwhelm society with their numbers. Immigrant bodies carried inside them microscopic time bombs. If their germplasm entered into the population, they'd permanently contaminate it with their inferior traits.

Scientific concerns about sexual relations between biologically distinct peoples had first spiked in the years after the Civil War. Presuming that the bonds of slavery had stymied relations between European Americans and the forced migrants from Africa they'd enslaved (they hadn't, though few would openly acknowledge it), scientists worried that the abolition of slavery might allow people of African and European descent to mix more freely. The crossing of biologically distinct subspecies, the Harvard University biologist Edward Murray East wrote, would "break apart those compatible physical and mental qualities which have established a smoothly operating whole in each race by hundreds of generations of natural selection." Anti-miscegenation laws that banned interracial sex and marriage, which had been passed in the 1860s, protected the nation from such an outcome. But no such laws protected the country's more advanced subspecies from the more primitive ones arriving daily on steamships from Russia, Poland, and elsewhere.

Ripley wasn't the only one raising the alarm. Leading eugenicists such as the Harvard zoologist Charles Davenport, founder of the Eugenics Record Office at Cold Spring Harbor Laboratory;

his managing director, Harry Laughlin; and top public health experts agreed.

The precise outcome of racial hybridization remained unclear. If people from tall races crossed with people from short races, some eugenicists worried, they could bear tall offspring with too-puny organs, or short offspring with grotesquely large organs. Their pairings might result in savage offspring who reverted to the ancient primitive type of one parent, like a domesticated plant that reverted back to wild type, losing the more evolved racial attributes of the other parent.

Americans could "rapidly become darker in pigmentation, smaller in stature, more mercurial, more attached to music and art," Davenport warned. They could become "more given to crimes of larceny, kidnapping, assault, murder, rape and sex-immorality."

Whatever the biological outcome, the hybrids would spell "absolute ruin" for American society, the physician Walter Ashby Plecker warned in an address to the American Public Health Association. His colleagues agreed, publishing a transcript of his remarks in the *American Journal of Public Health*.

As a big-game hunter, Grant had witnessed with sorrow the decline of the country's majestic large mammals. As he and Osborn absorbed the biological implications of immigration, they saw a similar process of displacement unfolding against their own kind.

"Miscegenation," Grant wrote, "is the first step toward extinction." Immigrants, by contaminating the nation with their inferior germplasm, would breed superior human subspecies into oblivion. As the genteel members of the Half-Moon Club furrowed their brows in their grand-palazzo-style clubhouse, outside on the streets of New York, the newcomers birthed a nation of hybrid monsters.

$\approx$

Most Americans in the years leading up to the First World War generally accepted that certain peoples—foreigners such as Asians and Africans, those deemed "feeble-minded"—were backward and undesirable and had to be kept at arm's length.

Congress had closed U.S. borders to people from China and to anyone judged to be suffering from lunacy or idiocy back in 1882. Dozens of states across the country prohibited "feeble-minded" people from getting married, for fear that their "feeble-minded" offspring would contaminate the populace. Some states even legalized their forced sterilization. "Society has no business to permit degenerates to reproduce their kind," President Theodore Roosevelt had written in a 1913 letter to Davenport. At the Bronx Zoo, crowds regularly gathered around the cage of Ota Benga, "most of the time roaring with laughter," as the *New York Times* reported.

But the finer points of race science and Weismannism escaped them. They sensed little biological danger emanating from the people from Mexico who easily traveled across the border into the United States, nor, for the most part, from people from Europe, who enjoyed nearly open access into the country as well.

While scientific elites detailed the biological menace of migration, popular culture embraced it. Hundreds of thousands cheered as workers erected the Statue of Liberty in New York Harbor in 1886, with the refugee advocate Emma Lazarus's sonnet, "Give me your tired, your poor / Your huddled masses yearning to breathe free," inscribed on a plaque at her feet. Across the city, so-called settlement houses strove to assimilate the newcomers, providing cooking classes, debating societies, and sewing instruction to help them shed their native customs and adopt American habits (eating creamed codfish and corn mush, for example, rather than the typical Mediterranean fare of meat, vegetables, and pasta, which late nineteenth-century American experts considered "overstimulating" and indigestible).

The 1908 musical *The Melting Pot* extolled immigrant assimilation. In the play, the main character, a Jewish refugee from Russia, falls in love with and marries a Christian refugee. He proclaims the United States as a nation of hybridization and amalgamation. "America is God's Crucible, the great Melting Pot where all the races of Europe are melting and re-forming!" he declares.

> Here you stand, good folk, think I, when I see them at Ellis Island, here you stand in your fifty groups, with your fifty languages and histories, and your fifty blood hatreds and rivalries. But you won't be long like that, brothers, for these are the fires of God you've come to—these are fires of God. A fig for your feuds and vendettas! Germans and Frenchmen, Irishmen and Englishmen, Jews and Russians—into the Crucible with you all! God is making the American.

President Roosevelt attended on opening night along with his cabinet secretaries. "Roosevelt watched the play enthusiastically," the *Times* reported, "at certain points shouting out at certain lines, 'That's all right!' while leaning forward in his box, and being the first to lead the applause at the end of the second act."

The positive economic and cultural impact of the new immigrants—and their ability to deliver winning votes in elections—impressed politicians far more than any potential biological impact they posed. For most pre–World War I American politicians, "the more the merrier," as one contemporary put it.

Plus, an early twentieth-century study commissioned by Congress—to this day, the largest study of immigration ever conducted in the United States—had discovered none of the biological hazards that so worried Osborn and Grant and the rest of the scientific establishment. A nine-member bipartisan commission had used all the latest social science techniques to look at how immigration influenced everything from crime rates and education to public health. The commission collected statistics on regional demands for labor and the proportion of foreign-born people in penal and charitable institutions. They investigated the conditions on immigrant ships, sending undercover investigators to report on the quality of the food. They analyzed how immigrants were received in U.S. communities. Did they join trade unions? How did unions accept them? Did their presence affect employment rates for native-born workers? What were their wages compared to those of other workers? What types of jobs did they take? Did they cause more accidents than native workers? Did their kids enroll in school? Could they speak English? What was the nature of their criminal tendencies? Their medical status? How often did they go insane? The crusading anthropologist Franz Boas—a prominent critic of race science—had even wrangled a few thousand dollars from the commission to measure the body dimensions of thousands of immigrant schoolchildren, to search for clues as to how the process of migration might have altered the shape of their bodies themselves.

The commission produced over twenty thousand pages of reports issued in forty-one volumes. Not only did the commissioners find no biohazards (or any other kind of hazard) associated with immigration, they suggested that scientists' depiction of foreign bodies as unalterably defective, the basis for their warnings about the biohazards of immigration, was mistaken.

Boas's study found that far from being permanently fixed by their unchanging germplasm, the body dimensions of immigrants and their children started to undergo a process of change as soon as they arrived in the United States. Exposure to new diets and environments transformed them physically, just as *The Melting Pot* had suggested.

# Prehistoric migrations to the Pacific

Early twentieth-century scientists considered the Pacific islands ideal locales to study the biological impact of migration, presupposing that people had started migrating there in modern times. This map shows the prehistoric waves of migration that peopled the Pacific islands, according to recent genetic and archaeological evidence.

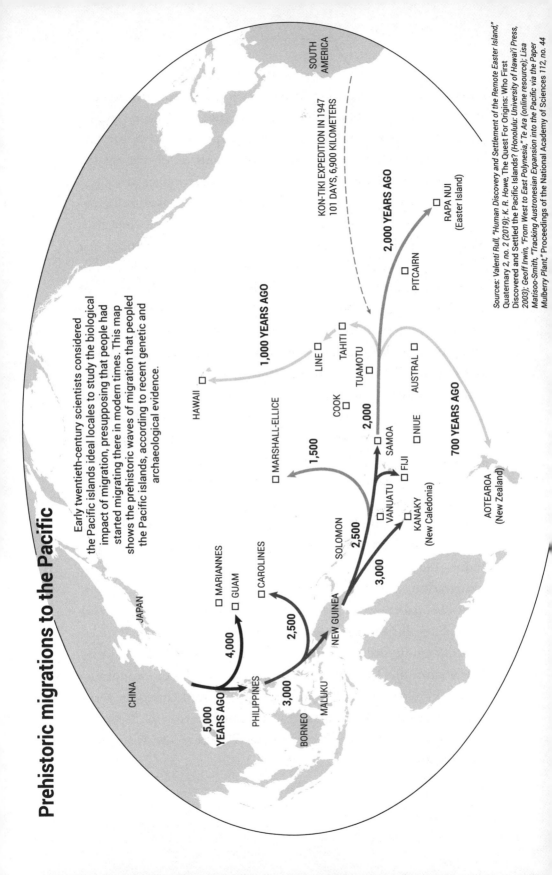

KON-TIKI EXPEDITION IN 1947
101 DAYS, 6,900 KILOMETERS

5,000 YEARS AGO

4,000

3,000

2,500

3,000

2,500

3,000

2,500

2,000

1,500

1,000 YEARS AGO

2,000

2,000 YEARS AGO

700 YEARS AGO

CHINA
JAPAN
MARIANNES
GUAM
CAROLINES
PHILIPPINES
BORNEO
MALUKU
NEW GUINEA
SOLOMON
VANUATU
KANAKY (New Caledonia)
FIJI
SAMOA
NIUE
AUSTRAL
AOTEAROA (New Zealand)
MARSHALL-ELLICE
HAWAII
COOK
LINE
TAHITI
TUAMOTU
PITCAIRN
RAPA NUI (Easter Island)
SOUTH AMERICA

Sources: Valentí Rull, "Human Discovery and Settlement of the Remote Easter Island," Quaternary 2, no. 2 (2019); K. R. Howe, The Quest For Origins: Who First Discovered and Settled the Pacific Islands? (Honolulu: University of Hawai'i Press, 2003); Geoff Irwin, "From West to East Polynesia," Te Ara (online resource); Lisa Matisoo-Smith, "Tracking Austronesian Expansion into the Pacific via the Paper Mulberry Plant," Proceedings of the National Academy of Sciences 112, no. 44

While not specifically designed to test the hypotheses of Weismannism and race science, all the commission's findings, in one way or another, flew directly in the face of its predictions. Grant privately scoffed at Boas's "silly" results. Most likely the immigrant children's harlot mothers had been having clandestine affairs with "real" Americans, endowing them with novel germplasm, he figured rudely. But the truth was, the commission had looked hard and found none of the effects that he, Osborn, and other top scientists fretted over.

Grant had hoped that the commission's findings would lead to laws to "keep out a great mass of worthless Jews and Syrians who are flooding our cities," as he put it. Instead, the report went nowhere, fizzling out in just a single piece of legislation that President William Howard Taft promptly vetoed.

~

The biohazard of immigration finally started to attract public attention as the country headed into the First World War.

In 1916 Grant published *The Passing of the Great Race,* in which he set out his ideas about the deep biological and historical origins of the racial hierarchy, and the dangers of upsetting it through migration. It slowly grew into a best seller. Roosevelt claimed to be so excited about the book, he would not just read it, he said, but "study it." Pulitzer Prize–winning journalists quoted the book in their articles, arguing, as Kenneth Roberts did in the *Saturday Evening Post,* that immigration would turn the American population into "a hybrid race of people as worthless and futile as the good-for-nothing mongrels of Central America and southeastern Europe." Hundreds of other books, aimed at general audiences, described the science behind the inferiority of non-European races.

At universities across the country, scientists taught courses on the biology of heredity. Between 1914 and 1928, the number of universities teaching eugenics jumped from 44 to 376, including top ones such as Harvard, Columbia, and Brown. At public events, popular-science-education groups such as the American Eugenics Society organized "Better Babies" contests and "Fitter Families for Future Firesides" competitions to raise awareness about the wonders of good germplasm. Filmgoers and readers absorbed Weismannism's central premises through human dramas and revisionist history. The Hollywood movie *The Black Stork*, for example, depicted the tale of a couple with mismatched germplasm who disregard warnings against having children, sorrowfully bear a "defective" child, and allow him to die.

Anti-German propaganda and anxieties about Communists in the wake of the 1917 Russian Revolution stoked Americans' anxieties about foreigners. New social science research on immigrants purported to reveal their biological backwardness as well. Administering newly developed intelligence tests on immigrants arriving at Ellis Island in 1917 showed that 83 percent of Jews, 80 percent of Hungarians, 79 percent of Italians, and 87 percent of Russians were "feeble-minded." A survey of the national origins of inmates of psychiatric institutions found that recent immigrants were disproportionately represented, comprising 13 percent of the general population but 19 percent of the insane population. "Insane aliens stream in steadily," the *New York Times* headlined its story on the findings. "One alien out of every 50 becomes a lunatic," *Harper's Weekly* claimed, while "the ratio among native Americans is one in 450." During the war, officials administered intelligence tests to nearly 2 million military recruits. They found that 89 percent of black soldiers qualified as "morons," and the intelligence of foreign-born peoples descended steadily from west to east, with English and Dutch people scoring the highest, and Russian, Italians, and Polish the lowest.

Methodological biases accounted for the findings, though few noticed at the time. The intelligence test was supposed to measure intellectual ability but in fact demanded answers to questions only people of a certain class and culture would know: the author of *Robinson Crusoe*, the Union commander at Mobile Bay, the product advertised by a character called Velvet Joe, what a first-class batting average was. (One of the test's administrators, Carl Brigham, went on to develop the first SAT.) Anyone on the margins of mainstream middle-class culture was destined to fail.

At Ellis Island, officials administered the culturally biased test to newcomers who spoke little English, faced a frightening and confusing entry process, and had just endured arduous, days-long journeys under harsh conditions. All were exhausted, which would have impaired their capacity to ace any kind of test, even if it hadn't been biased against them. Similarly, the study on immigrants in insane asylums hadn't corrected for age distribution. The fact that the immigrant population, as a whole, skewed younger than the native-born one, explained their disproportionate representation entirely.

~

Even as flawed research piled up showcasing the biological inferiority of immigrants, the most urgent threat they posed to the nation—the catastrophe of racial hybridization—remained mostly theoretical.

Conflicting evidence had emerged. Subspecies theory predicted that hybrids would be less fertile than pure types, or even sterile like mules, but studies on the progeny of mixed-race couples suggested just the opposite. Boas's study of 577 Native American women found that they produced 5.9 children on average, while the 141 mixed-race women he studied averaged 7.9 children. Even the German anthropologist Eugen Fischer, whose research would later form the scientific basis for the Nazi Party's Nuremberg Laws, noted that the racial hybrids he'd studied—the children of Boer colonists and "Hottentots" in southwestern Africa—seemed perfectly fertile. The Swedish physician Herman Lundborg scrutinized photographs and measured the faces of the progeny of Lapps, Finns, and Swedes. He found, to his surprise, that the hybrids seemed taller, stronger, and more graceful than their pure-race ancestors. Boas found that mixed-race children grew taller than nonmixed children.

Still, innuendo and speculation about hybridization abounded. Many scientists felt certain that so-called mulattoes, people born to one black and one white parent, showed signs of dysfunctionality. They had "irregular dentations," Davenport claimed. They were a "nuisance to others," he wrote, because they had inherited ambition from their biologically superior white parents but "intellectual inadequacy" from their black ones. Mulattoes had no sagittal sutures in their skulls, preventing lateral expansion, the president of the Philadelphia County Medical Society asserted. Look at Haiti, other scientists said, where a 1791 revolution against French colonial rule had led to what Davenport's colleague Harry Laughlin called a "reversion to African barbarism." The lurid gossip they'd heard about the island, of cannibalism and worse, probably stemmed from its large mulatto population, they said.

Research on the pressing question presented a range of practical difficulties. It wasn't as if scientists could cross different races of peoples the way they crossed different breeds of rabbits and dogs and then evaluate their fitness. They had to rely on data from mixed-race couplings that transpired naturally in society. First, they had to find mixed-race people. Given the social disdain directed at them, that wasn't easy. Boas had resorted to sending grad students such as the anthropologist and writer Zora Neale Hurston to stand on street corners in Harlem with a pair of calipers in her hand in the hopes of flagging down a passing "mulatto" to measure.

Reconstructing a hybrid's racial history all the way back to "pure" race ancestors posed another challenge. Few subjects knew their ancestors' history, and even if they did, they didn't care to share it. Records such as those kept by churches were similarly unreliable, often listing as parents people other than biological fathers.

Finally, detecting the physical degeneration of hybrid bodies required finesse. In mixed-race experiments in rabbits, say, scientists could evaluate the fitness of hybrids by counting the number of babies they produced, or noting whether their ears stood upright or drooped. But detecting the physical degeneration of racial hybrids required taking dozens of detailed measurements. It wasn't as if the hybrid products of migration were *obviously* monstrous. Detecting their monstrosity required paying attention to detail.

Researchers aspired to conduct studies in places where distinct human subspecies mixed openly with no social stigma. The best place, many race scientists agreed, were Pacific islands such as Hawaii. The United States had annexed the lush volcanic islands of Hawaii in 1898. Over the course of the following decades, waves of migrants from the United States, Japan, China, and elsewhere had transformed the local population. White people married Hawaiian people, who married Chinese people, who in turn married Japanese people. Their mixed-race children married other mixed-race children. The promiscuously hybridizing Pacific Islands provided "a kind of laboratory," the *Times* noted, "in which nature may be watched as she performs the miracle of welding alien types." There was no social stigma about it; census records and death certificates tracked the whole process. "Possibly no equal area in the world presents more interesting racial aspects," the public health statistician Frederick Hoffman enthused.

Osborn's Museum of Natural History often sponsored scientific expeditions. It sent explorers to the North Pole, to uncharted regions of Siberia, to Outer Mongolia, and to the jungles of equatorial Africa. With the biology of race mixing being one of the most pressing scientific and political issues facing the nation, the museum sponsored a new expedition to produce a definitive study on the subject. In 1920 it sent Louis Sullivan, a PhD student from Columbia University, to Hawaii to conduct the necessary studies.

Osborn was in the midst of organizing an important international scientific conference on Weismannism and race science, during which he hoped to sway public opinion against immigration once and for all. The biology of racial hybridization would be high on the agenda. With any luck, Sullivan would produce definitive results in time for him to make the case.

~

In late September 1921, leading scientists from the United States, Europe, and elsewhere descended on New York City for the Second International Congress of Eugenics. The entire fourth floor of the American Museum of Natural History had been cleared out for the gathering. The inventor and scientist Alexander Graham Bell attended, as did one of Darwin's sons, Major Leonard Darwin, president of the Royal Geographical Society, among other scientific luminaries of the day.

During his opening address, Osborn explained the political aims of the conference. Hundreds of papers and exhibits would showcase the latest findings in race science and Weismannism and demonstrate the scientific urgency of ending immigration and race mixing, he explained. "We are engaged in a serious struggle," he told the assembled attendees, "to maintain our historic . . . institutions through barring the entrance to those who are unfit."

Over one hundred exhibits were displayed in the museum's exhibit hall. There were enlarged maps from Madison Grant's best-selling book and a ghoulish display featuring plaster casts of fetuses that claimed to show that African American fetuses had smaller brains than white ones. Another one contrasted enlarged photographs of the brains of criminals and those of the "feeble-minded." Charts showcased the fecundity of immigrants.

The Census Bureau provided several diagrams intended to underline certain points about race and migration, such as a chart comparing the number of white people and the number of nonwhite people in insane asylums. Davenport offered ten pedigree charts on the "Inheritance of Genius and Talent in American families," featuring the Perry family of naval officers, the Jefferson family of actors, the Agassiz family of scientists, and others, and a display of sixty-one photographs of the "racial types" among recent immigrants at Ellis Island called "Carriers of the Germ-Plasm of the Future American Population."

For a week, the gathered attendees listened to lectures. Major Leonard Darwin argued that "the inborn qualities of civilized communities are deteriorating." Scientists from Cold Spring Harbor explained how musical, literary, and artistic skills were inherited biologically; others opined on whether it was possible to breed geniuses, why redheaded people "dislike one another," and why tall men choose short wives and short men choose tall wives. Scientists explained why superior intelligence was five times more common in children of parents with superior social status, and how "democratically minded persons" mouthing "benevolent platitudes" resisted these scientific facts of nature.

But the most critical scientific question at the congress concerned the fundamental biological problem posed by immigration—the biology of race mixing. The attendees eagerly awaited results from Hawaii, where Sullivan felt certain he was on to something.

"I'm head over heels in the Polynesian problem," Sullivan had written to his sponsors. Soon, he hoped, he'd discover "the ultimate solution of race relationships." He'd measured nearly eleven thousand Hawaiians and analyzed more than three hundred skulls. He'd measured the body parts of mixed-race children. He'd sampled their blood. He'd taken hair samples. He shot photographs of his research subjects, clothed, and then when Osborn demanded that he needed them to be posed naked for his exhibit, unclothed. Race mixing "from the standpoint of the Whites or Chinese," he wrote confidently in private correspondence, "is a failure of course." But to figure it out for sure, he'd need more time to sift through the mountains of data, scribbled on index cards, overwhelming in volume, and yet stubbornly cryptic. While Sullivan couldn't attend the conference, he'd sent photographs, face casts, and charts, which curators assembled into a display on the "race problem in Hawaii," along with statistics and photographs contrasting "pure" Hawaiians, Chinese, Japanese, and Portuguese people with the "mixtures" they'd sired.

A colleague appeared at the conference to present some preliminary evidence from Hawaii, reassuring the audience that Sullivan's "authoritative account" was forthcoming. In the meantime, conference speakers presented a mix of mostly inconclusive studies on racial hybridization. A scientist from the Carnegie Institution presented his work on mixed-race mice, which had found that the hybrid mice exhibited greater strength and adeptness at mazes than their pure-race parents. Another speaker, delivering a paper on intermarriage, pointed out that countries with more racially hybridized populations such as the United States and England demonstrated a higher stage of "mental evolution," as he put it, than those with less racially hybrid populations, such as countries in Central Asia and Africa.

Osborn's favorite, though, was that of the Norwegian biologist Jon Alfred Mjøen, who reported on his study of racial hybrids in humans and rodents. Osborn praised Mjøen's paper as a "splendid contribution." Mjøen had mated several different breeds of rabbits, then mated their hybrid offspring to each other, and so on for five generations, producing a population of degenerated, infertile, sickly rabbits. Mortality rose from 11 percent in the first generation to 38 percent by the fifth generation, he told conference attendees, by which time the rabbits were so impaired that they wouldn't mate and some of them had one upright ear and one pendant ear.

The rabbits had been damaged by being isolated in a small group and mated with their own relatives for five generations. But Mjøen interpreted the contours of their bodies—their strangely divergent ears, for example—as the effects of racial hybridization. And those were just the most obvious of the ill effects he suspected lurked deeper within their furry bodies. "Why should only the ears be affected?" he asked his audience. "We ought to be suspicious in regard to every organ: heart, lungs, kidneys, bones. In fact, we must be suspicious in regard to the whole organism of the hybrid, when we see this most striking disharmony."

Mjøen claimed to have found similar results in humans. Like Linnaeus years earlier, he studied Laplanders, more specifically Lapp-Norwegian hybrid people, whom he evaluated for physical and mental deficiencies. Most of the hybrids, he found, exhibited what he called the "M.B. type," which stood for "Mang-Lende balance," or "want of balance." They were good-natured and willing but unbalanced and unreliable, he decided, with the main symptoms of their hybridized condition being stealing, lying, and drinking. To illustrate his point, he displayed a photograph of three boys, sitting on a rumpled blanket in front of a wooden shack, whom he described as mentally degenerated racial hybrids. To modern eyes, it is obviously and crudely doctored, though Mjøen did not present it as such.

Mjøen admitted that his research provided no definitive proof of the danger of racial hybridization. That would have to wait for Sullivan's results. But given the apparent risks of racial hybridization, he told his listeners, prudent policy makers should aim to "nourish and develop a strong and healthy race instinct," even in the absence of definitive proof. Language instruction and other assimilation services aimed at immigrants should be shut down, for they built bridges between the races, the results of which "we will deplore and regret when it is too late."

When the week came to a close, the conference attendees departed, spending Sunday on a special excursion to the Bronx Zoo. Afterward Osborn arranged for the conference's exhibits to be sent to Washington, D.C., to be displayed in the halls of the Capitol.

Meanwhile Grant, Osborn, and other scientists who'd participated in the conference formed a new organization to distill the latest scientific findings into concrete policy. With Grant as their chair, they penned a new immigration law that one of their allies, Representative Albert Johnson, would introduce into Congress.

∼

As chair of the House Committee on Immigration and Naturalization, Johnson had been instrumental in raising awareness of the biology of migration among his fellow lawmakers. He'd appointed Davenport's colleague Harry Laughlin as an "expert eugenics agent" to the committee, arranging for him to testify about how "queer, alien, mongrelized people" had to be kept out of the country, and calling his testimony, which was published in pamphlet form, "one of the most valuable documents ever put out by a Committee of Congress." By the time Johnson introduced the immigration bill drafted by Grant's committee in late 1923, he felt certain that scientific questions about immigration had "been settled in the minds of members of the House and Senate," as he wrote in his private correspondence. Grant agreed. "You have the country behind you and a most popular cause," Grant assured Johnson.

He had the newly installed president behind him, too. Calvin Coolidge had ascended to the presidency just months earlier, after the unexpected death of President Warren Harding. In a 1921 *Good Housekeeping* article, Coolidge had written about "biological laws" that "tell us that certain divergent people will not mix or blend."

When debate on the bill opened, it was clear that the political establishment's views on the value of assimilating newcomers had shifted dramatically since the heady days of *The Melting Pot*. In the months leading up to the vote, Grant's book was selling so fast, it had to be reprinted every six months.

The blood of America, one member of Congress proclaimed, had to be "kept pure."

"We are a different race," another added.

The immigrants "will vitiate our population."

Of the foreign-born people already in the country, nearly half were "inferior or very inferior" according to intelligence tests, a congressmember pointed out. "We can readily see the effect on the American people of this steady incursion of individuals of low mental capacity." Contaminated by inferior foreign biomatter, future generations would be permanently diminished.

Dismissing the complaints of steamship companies, immigration advocates, and a sprinkling of pro-immigrant congressmembers— such as one who eloquently dubbed Grant's ideas about superior races "senseless jargon," "pompous jumble," and "dogmatic piffle"—the bill passed with large majorities in both the House of Representatives and the Senate.

Laws that temporarily and partially restricted immigration had already been passed into law during the 1914–18 war. The bill that Grant and Osborn's committee wrote would expand and make permanent those restrictions, war or no war. President Coolidge readily signed the bill. "America," he proclaimed, "must be kept American."

~

Sullivan never finished his "authoritative" study of race mixing. He fell ill with tuberculosis and had to abandon the study. He died in 1925. The American Museum of Natural History sent the Harvard anthropologist Harry Shapiro to carry on his unfinished work.

If anyone could complete the research, Shapiro could. His focus was ruthless. Once, when Shapiro heard of a recently buried skull in a cemetery, he secreted off to the site at night, stealing it and hiding it in his laundry. Another time, when he'd heard of a burial ground high in the Tahitian mountains, he and a colleague set off on a treacherous grave-digging excursion to secretly dig them up; they gingerly descended the steep, jungled slope with their axes in hand and their knapsacks heavy with stolen skulls on their backs. "I had the continuous impression of being ready to go headlong at any moment," he recalled. "When we finally reached bottom I could hardly see through my glasses," which were coated with a film of dirt and sweat. The researchers triumphantly sent the bones to the museum, celebrating their success that night with a steak dinner.

And yet as his research progressed, Shapiro's faith in the study's fundamental premises wavered. His task was to find evidence of the dangers of migration's most common result, miscegenation. And yet somehow he felt drawn to his subjects, not just by scientific curiosity but also by passion and desire. He lived in local Hawaiians' houses and accepted their gifts of shells, baskets, and food, their friendly kisses and handshakes. He admired their "liquid eyes" and "soft, languorous expression[s]." At some point, Shapiro starting having sex with his subjects.

His confusion deepening, Shapiro decided to leave Hawaii and delve into the question elsewhere. Pitcairn Island, he believed, was an even better research site for race-mixing studies than Hawaii. In 1789 nine English mutineers of the Royal Navy vessel *Bounty* had settled the island with a group of Polynesian women, creating a hybrid population out of two racially distinct groups. But Pitcairn, a tiny speck in the vast Pacific, was not easy to get to. Shapiro attempted to reach it in 1923, traveling on a ship bound from Panama to New Zealand in hopes of jumping off board when the boat neared Pitcairn. But as the ship approached the island, a tropical storm forced the captain to alter course, foiling Shapiro's plan to lower himself onto a smaller boat and secretly row to shore.

Shapiro finally arrived in Pitcairn in 1934. Here he hoped he'd find "definite indications" of the degeneration caused by racial hybridity. With the two parent races of the Pitcairn Islanders so distinctive, he expected to find a wide range of diverse effects in their hybrid descendants: changes in their health status, the diseases they suffered, their height, their skin, and their fertility.

But when he met the islanders, they were not monsters. He found them to be "more like a group of Englishmen dockworkers," he wrote, "with ugly knobby hands and feet rough and calloused by labor." He took detailed measurements of their height, the length and width of their heads, the distance between their nasal septa and the nasal routes, and the thickness of their lips. He noted the color of their eyes, hair, and skin. But even with dozens of measurements, he could find no evidence that the Pitcairn Islanders had developed into anything other than normal humans.

They were physically robust, suffered few diseases, exhibited generally average intelligence, and bore plenty of healthy babies.

"The Pitcairn Islanders," he reported, "show no ill effects of several generations of intermarriage. They are taller, and at least in some respects appear to be better developed physically, than either the English or the Polynesian races."

The only thing wrong with them he could find was their bad teeth.

Other researchers' studies on racial hybridization similarly fizzled. Davenport published his study on race mixing in Jamaica in 1929. He hadn't been able to find much difference between black and white people and their mixed-race offspring there. "Physically there is little to choose between the three groups," he admitted. The most dire effect he could find in the hybrids was intelligence he personally deemed "mediocre," and the fact that a few of them had "the long legs of the Negro and the short arms of the white," which he claimed put them "at a disadvantage in picking up things off the ground." (Even that claim was exaggerated, as a colleague found later when he reanalyzed Davenport's data: the arms of the mixed-race subjects could reach—at most—one centimeter less than those of their pure-race parents.)

Shapiro returned to the United States a changed man. He'd spent much of his career attempting to document the dangerous biological effects of hybridization. He realized he'd been chasing a mirage.

When he wrote up his notes on his transformative trip to Pitcairn into a book, he devoted just a sliver of his attention to the biological effects of racial hybridization, focusing instead on the novel cultural traditions that the migrant peoples and the natives had forged. He went on to conduct a groundbreaking study that showed how migration—not fixed characteristics of race or subspecies—influenced our bodies. In the study, he compared Japanese migrants to Hawaii, their Hawaiian-born children, and their nonmigrating relatives in Japan. Just as Boas had found decades earlier, the landscapes into which they'd migrated had shaped their bodies: the Japanese migrants' children were taller than those of their nonmigrating relatives in Japan. Their shared "race" had nothing to do with it.

"Man emerges as a dynamic organism," Shapiro wrote, "which under certain circumstances is capable of very substantial changes within a single generation." Shaped by a long history of migration, the human body was not rigidly constrained to any one place or type, subspecies or race, dictated robotically by germplasm or anything else.

By the mid-1930s Shapiro had renounced the scientific presumptions that had driven a generation of scientists, federal immigration policy, and his own years of research. The mixing of peoples from different places posed no peril. Just the opposite: by injecting change and innovation into cultural practices, migration was "an integral factor in the history of human civilization," as one of his biographers put it. But by the time race-mixing biology imploded, it was too late.

~

Convinced of the biological hazard posed by immigration, Congress had passed the immigration law penned by Grant's eugenics committee. Under the Johnson-Reed Act, strict new quotas protected the nation from those whom scientists deemed racial inferiors. Under the act's terms, over 80 percent of the annual quota of immigrants would be reserved for people from western and northern Europe. The vast majority of nonwhite immigrants and people from eastern and southern Europe would be barred entry. A newly formed Border Patrol force would enforce the rules at the border.

The flow of immigrants into the United States plunged from over 800,000 in 1921 to 280,000 in 1929 and fewer than 100,000 a year after that. The spigot tightened, the flow of newcomers slowed to a trickle. The Ellis Island immigration station shut down in 1954. Its services were no longer needed. The era of nearly open borders with Europe was over.

Word of the newly emerged Fortress America and the scientific principles on which it was based traveled across the globe. A pro-Nazi publishing company in Germany published Grant's book in 1925. Adolf Hitler read it while stuck in a Bavarian jail. "The book is my bible," he told Grant in a letter, as he started to envision his own program of ridding the nation of those deemed outsiders.

When his genocidal regime forced masses of Jews and other unwelcome outsiders to flee the country, the United States did not waver from its commitment to closed borders. "We must ignore the tears of sobbing sentimentalists and internationalists," a member of the House Committee on Immigration said, and "permanently close, lock and bar the gates of our country to new immigration waves and then throw the keys away." In polls conducted in the late 1930s, two-thirds of Americans said they agreed.

In February 1939 a bipartisan bill granting twenty thousand Jewish children asylum from the Nazi regime was introduced into Congress. President Franklin Delano Roosevelt pointedly took no stand; his wife, Eleanor Roosevelt, didn't either. Anti-immigrant congressmembers killed the bill. Twenty thousand "charming children would all too soon grow into 20,000 ugly adults," one advocate, the wife of the U.S. immigration commissioner, testified.

A few months later an ocean liner carrying more than nine hundred terrified asylum seekers from Germany arrived in Miami. U.S. officials refused to let the ship dock, calling the Coast Guard in as reinforcements. For days, the ship circled off Florida's coast, its passengers sobbing on the balconies, until the captain finally steered it back across the Atlantic to war-torn Europe. Some of the ship's passengers made their way to Britain. Most ended up in the Netherlands, Belgium, and France, where they soon faced Nazi occupation. Over 250 died in the Holocaust.

# 5

# THE SUICIDAL ZOMBIE MIGRANT

W hile Grant and Osborn incited alarm about the disorder caused by human migrants crossing geographical and biological borders, scientists in Britain pondered the challenges faced by growing populations constrained within those borders. In their vision of nature's order, migrants played a macabre role. The most fitting end of migrant journeys, leading scientists said, was death.

The theory of migrant death began in the Arctic, where scientists first encountered local stories about furry Arctic-dwelling rodents called lemmings. In 1924 Charles Sutherland Elton was a twenty-four-year-old undergraduate in zoology at Oxford. Hired as an assistant on a series of expeditions to Spitsbergen, a then-uninhabited Arctic island halfway between Norway and the North Pole, he had helped conduct an ecological survey of the polar bears, walruses, reindeer, lemmings, and other Arctic creatures that roamed the snowy vistas.

The expedition provided a valuable opportunity for the ambitious young scientist. For weeks, Elton shared adventures and intimate living spaces with some of Oxford's most accomplished scientists: Aldous Huxley's brother, the geneticist Julian Huxley; the sociologist Alexander Carr-Saunders, who'd later become director of the London School of Economics; the Rhodes scholar Howard Florey. But while Spitsbergen provided the backdrop for Carr-Saunders's writing of his magnum opus—Elton found it "full of exciting ideas of a general nature"—and Florey would go on to develop penicillin, opportunities for Elton to use the survey in Spitsbergen to catapult his own career appeared slim. Elton spied no unusual animals nor any never-before-seen behaviors that might help him make his mark in natural history. At one low point, he fell through the ice into a lake, whose waters submerged his body up to his neck.

Elton did not happen upon his chance until after he'd left the island. While sailing back to Oxford, he and the others stopped briefly in the northern Norwegian city of Tromsø. Elton ventured into the town, past its centuries-old wooden houses nightly illuminated by the Northern Lights, eventually finding himself in a small bookstore. Browsing, he encountered a book called *Norway's Mammals* by the Norwegian zoologist Robert Collett, published in 1895. He pulled it off the shelf and paged through it. It was the only book in the store on natural history, his subject of interest, but it was written entirely in Norwegian. He couldn't read a word. He'd likely have pushed the book back onto the shelf had it not fallen open to reveal a few pages that included something Elton could read: several charts listing columns of numbers.

Elton had no idea what the numbers meant, so he brought the book over to the shopkeeper to ask. "Peak lemming years," he said.

~

Elton, though an accomplished student of zoology, had little interest in traditional natural history. To him, natural history consisted of idiosyncratic portraits of individual creatures, penned by eccentric animal-loving observers. It had little relevance to the urgent questions of the day, which revolved around issues like famine, war, and epidemics of pests and pathogens. He wanted to revolutionize natural history into something more like what early twentieth-century physics and chemistry had become: muscular, hard-hitting, and capable of rendering practical, economy-changing insights.

Instead of looking at the behaviors of individual animals, Elton thought, zoologists should study the "sociology and economics of animals," that is, how whole populations behaved, in relation to one another and to the environment. And so while the descriptions of Norway's mammals in Collett's book held little appeal for him, the numbers in Collett's charts did. He could see that the "peak lemming years" occurred intermittently, which meant that the numbers of lemmings rose and fell cyclically over time.

These changes in population size mystified zoologists at the time. Their confusion stemmed from their understanding that nature was divided into biologically discrete habitats. Naturalists described the places that wild species lived in as "niches," from the Middle French word *nicher*, "to nest." The word had originally referred to the recess in a wall carved out to nestle a statue. Zoologists imagined each wild species' niche to be similarly specific and unique, carved out to fit the one species that occupied it. Each species lived in its own piece of nature with biological borders drawn around it.

That conception led to a paradox, though. Scientists understood niches by studying experimental versions in their labs. Since a niche was an enclosed space equipped with the necessities of life for its species, it could be easily replicated in a laboratory, for example by establishing a colony of yeast cells in a test-tube filled with sugar water. But real-world niches didn't behave the way experimental niches suggested they should. In lab experiments, the size of a yeast colony in its test-tube niche would rise and fall in direct relation to the volume of sugar in the tube. If scientists kept adding sugar, the yeast would keep growing. If they stopped, the yeast would stop.

If wild species lived in closed-border niches, as scientists thought they did, then the size of their populations should similarly expand and contract in direct relation to the availability of food and water. But that's not what zoologists saw in the wild. Animal populations did not grow until their food supply collapsed and they starved to death. Their numbers rose to a certain point, and then as if reaching some invisible ceiling, they started to decline again, in an endless series of cyclical spikes and falls. The availability of food and shelter made no difference. It was as mysterious as if yeast cells in a fully fueled test tube grew for a few days, then declined for a few days, then grew again.

Linnaeus had known about the conundrum of this so-called population cycle. He figured it had something to do with God. By Elton's time, zoologists had ruled out divine intervention. But if God couldn't explain it, neither could any of the other external factors zoologists looked at: food supplies, environmental disruption, predators, disease. It seemed as if some invisible factor X must secretly regulate population growth, like a finger pressing on a scale. But what was it?

The mysterious cycling of populations was no charming eccentricity of animal behavior. It was a phenomenon with great economic import. During low points in the population cycles of fur-bearing animals such as foxes, for example, hunters went hungry and the price of fur spiked; during the high points in population cycles of voles and locusts, the abundant creatures destroyed lucrative logging areas and agricultural fields. But with little scientific understanding of the factors that shaped population cycles, they could be neither predicted nor controlled.

Elton was young and ambitious. Hadn't Einstein been just twenty-six when he revolutionized physics during his *annus mirabilis* in 1905? If Collett's book allowed Elton to pinpoint how and why lemming populations rose and fell, he could pinpoint factor X and unlock an enduring mystery, too. Perhaps he could even distill the phenomenon of population cycles into a mathematical formula. He'd turn musty old natural history into a sturdy, quantitative science. He might even be able to predict and control the rise and fall of animal populations. Powerful companies such as the Hudson's Bay Company and British Petroleum, among others, would surely be interested in funding such inquiries.

He bought the book. When he got back to Oxford, he acquired a Norwegian-English dictionary and eked out a crude, word-for-word translation.

~

From its pages, Elton learned of the strange and mysterious behavior of the lemmings.

The book described reports of lemmings gathering together in great masses and marching toward the Arctic cliffs, where they flung themselves into the sea. One observer named Duppa Crotch who spent his summers in Norway had observed the phenomenon more than once. His report appeared in an 1891 issue of *Nature*. When he saw the lemmings in the water, he rowed his boat toward them to block their way. They swam by him determinedly, he reported, making a beeline to their watery graves. "I know nothing more striking in natural history," he wrote.

In 1888 a mass of lemmings formed "until the whole land was black with them," then started "moving seaward on a 10 mile front" that took four days to pass by. "They kept on over the sea ice, finally leaping into the water and swimming ashore until drowned," another observer wrote. Nineteenth-century sailors claimed to have seen millions of lemmings flailing in Norway's deep, narrow fjords. "So many swam out into the inner parts of Trondheim Fjord," one sailor recounted, "that a steamer took a quarter of an hour to pass through them." The mass migration ended badly for them. Scores of lemming carcasses had been encountered, flung across the surface of iced-over lakes, frozen to death.

What did it all mean? According to ancient Laplander legend, lemmings originated in the heavenly mountains and appeared suddenly when they rained down from the sky. Then they gathered in flocks searching for a way to return. Some said they were poisonous to the touch. For Crotch, the lemmings' migration into the sea suggested an ancient Atlantis. Crotch guessed that "blind and sometimes even prejudicial inheritance of previously acquired experience" caused lemmings to migrate into the sea. Perhaps, he speculated, their destination had once been dry land: an Atlantis hidden somewhere in northern Norway.

Collett's stories struck Elton differently. Perhaps, he speculated, the lemmings headed out to sea not because they were trying to get somewhere, but for the opposite reason: because they knew it would get them nowhere. It was a behavior similar to one that Elton's mentor Alexander Carr-Saunders had described in his best-selling book, which Elton had "devoured." On the Pacific island of Funafuti, Carr-Saunders had written, people ritually murdered every other baby born until each woman had four living children, after which every baby they bore also would be ritually murdered. This cultural practice, Carr-Saunders wrote, formed a crude but effective form of population regulation. (Sentimental outsiders would ruin it if they interfered, he warned, setting off a population explosion.)

By migrating to their certain deaths, the lemmings achieved the same result as the Funafuti. They culled the population, protecting it from the calamity of straining the limits of their food supply. Perhaps, Elton speculated, the suicidal lemming migration was no error or artifact but had emerged and continued to persist for that precise reason. Lemming populations expanded to a certain point, after which they committed mass suicide by migration, causing a consequent decline. That would explain why their mysterious population fluctuations did not coincide with famines or disasters.

Elton wrote up his novel spin on Collett's findings into a scientific paper, which appeared in the *British Journal of Experimental Biology* in 1924. The paper started out with a fairly neutral description of the phenomenon. "For many years," Elton explained, "the lemmings have periodically forced themselves upon public attention in southern Norway by migrating down in swarms into the lowland in autumn, and in many cases marching with great speed and determination into the sea, in attempting to swim across which they perish," he explained. As the son of a literary scholar and a children's book author—and the future husband of a poet— Elton could not resist the temptation to wax lyrical, even though he'd never seen a lemming, let alone a lemming migration. He wrote of lemmings "ecstatically throwing themselves over the ends of railway bridges," making a "bee-line across crowded traffic oblivious to danger," and the sea "strewn with dead lemmings like leaves on the ground after a storm."

Such colorful descriptions sparked the attention of his fellow zoologists. Even better, Elton had a tidy explanation for why the lemmings did it, one that shed light on the broader and economically urgent question of the origins of population fluctuations. "The phenomenon," he explained, "is analogous to infanticide among human beings . . . the immediate cause of the migration is overpopulation."

Elton had discovered the mysterious factor X that explained why animal populations rose and fell in cycles: it was a secret drive to regulate the size of one's population. Elton's paper, hailed as "visionary" and "seminal," renewed zoologists' interest in the issue of population fluctuations. It became "one of the cornerstones of contemporary ecology," as a 2001 paper in *Biological Reviews* put it. Oxford established a new research institute, the Bureau of Animal Population, installing Elton as its director. Scientists organized conferences focused entirely on population cycles, and across Europe and the United States they conducted lab experiments, performed field research, and searched for mathematical formulas to describe and explain animals' drive to regulate their own numbers.

Biologists discovered manifestations of this secret drive for suicidal migration in other species besides lemmings. The University of Michigan zoologist Marston Bates, for example, wrote of "mass suicide" committed by South American butterflies. "The pressure of built-up numbers seems to result in an explosive migration into new areas, where the migrating individuals die," he explained. "It is a sort of mass suicide." He had watched millions of South American butterflies fly out to sea "to certain death," a result of the "balance of nature." Scientists speculated about shoals of fish purposely dashing themselves to death on the hulls of boats, and suicidal whales beaching themselves on shore. Perhaps their self-destructive impulses, too, were compelled by their awareness of the size of their own populations, and a concomitant drive to sacrifice themselves for the good of their brethren.

The fact that established and highly trained scientists readily accepted the supposition that people purposefully murdered their children and that wild animals purposefully killed themselves is striking. They had faith in their conception of a nonmigratory, closed-border world. In fact, species were not confined to niches with impassable borders around them, like yeast trapped in a glass-walled test tube. Individuals moved into and out of populations.

And the environments within habitats were dynamic, too, which individuals responded to in idiosyncratic ways. Some did well at some times; others less so at other times. The size of a population rose and fell because both the composition of the population and the environments it lived under continually changed.

The population cycles that scientists pondered appeared paradoxical only because they didn't know that the borders around habitats were permeable and allowed migration to flourish. And they accepted the idea of suicidal migrations because Darwin's theory of natural selection was not widely accepted at the time. It allowed for no mechanism by which suicidal migrations could evolve. The traits of individuals who successfully reared more young than others dominated populations, not the traits of those who purposely destroyed themselves. If suicidal lemmings did emerge, they'd kill themselves and be replaced by lemmings who didn't recklessly leap from cliffs. Repeated acts of mass suicide, in other words, could not exist in nature.

But for zoologists at the time, the idea that uniform populations enclosed within niches migrated as a form of population control made sense. Twentieth-century zoologists imagined migration, in reality a vector of life-giving biological and cultural diversity that ecosystems and societies depended on, as a vector of death.

∾

For Elton, as for Linnaeus, the conviction that drove his antimigrant ideas concerned the past. For him, nature had always existed in stasis. Geography was immortal. The "principal masses of land and water," he wrote, "have existed mainly in their present shapes throughout all ages." Over time the unchanging landscape had been populated by wild creatures, each species establishing itself in its own niche. "Nearly all animals," Elton explained in one of his books, had become "more or less specialised for life in a narrow range of environmental conditions" through eons of habitation in their unique niches.

This conception of the past conformed to popular customs that elevated the native over the newcomer. The idea that plants and animals already resident in a place enjoy a special and privileged relationship to their habitats found expression across society. In museums, curators explained the specimens in their collections with little information other than their country of origin, as if that detail alone explained everything viewers might want to know. In English common law, people who lived in the countries of their birth enjoyed special rights of automatic citizenship, a provision known as *jus soli*, Latin for "right of the soil." The underlying idea about history necessarily turned migrants, whether they introduced defective germplasm or not, into ecological troublemakers.

In Elton's vision of nature, there was no excess capacity for newcomers, no "extra" niches. The most famous experiments proving the point had been conducted in the early 1930s. In 1932 the Russian biologist Georgii Frantsevich Gause introduced into a test tube filled with sugary liquid two different species of yeast: one called *Saccharomyces cerevisiae*, and the other, *Schizosaccharomyces kefir*. Every so often he shook the little tubes to ensure their contents remained fully mixed. He added nutrients for them to feed on. He refreshed them with water. He nourished the test tube with plenty of food to sustain both yeast species—but trapped inside the same tube, the two would have to share.

At first, populations of both species of yeast, fattened on the continuously replenished nutrients available to them, grew. But then, despite the abundance of available food and water, one yeast type started to suffer. Its numbers fell. Soon, as the population of its rival yeast type strengthened, the population of the declining yeast crashed, poisoned by all the ethyl alcohol in the waste of its tubemate. The phenomenon became known as "competitive exclusion" or, more simply, Gause's Law.

According to Gause's Law, there is no such thing as what we might colloquially call "sharing." Regardless of the abundance of resources, it is biologically impossible for two species to share the same niche. Either the newcomer or the native will be destroyed, poisoned into extinction like an alcohol-sensitive yeast in a glass test tube.

Years of experimentation and mathematical modeling confirmed Gause's findings. In part, that was due to the ease with which negative results could be dismissed. Biologists would dump two species with similar characteristics into the same location. Sometimes one would flourish while the other suffered, in which case they'd conclude that Gause's Law had been proved. Other times both would flourish, but instead of concluding that two similar species could in fact share the same niche, they'd instead claim that the two species must have been dissimilar, with some as-yet-undiscovered ecological difference between them. That is, they survived together because they didn't, in fact, share the same ecological niche.

In the belief that nature was in essence "filled up," experts in the United States and Britain started to target wild migrants as dangerous intruders. The arrival of newcomers, for them, signaled the certain demise of natives, as Gause's Law made clear. And according to Elton, migratory movements had no positive ecological function. Migrants embarked on their journeys in futile attempts to flee, not to arrive anywhere, Elton said. "Many animals migrate on a large scale," he wrote, "in order to get away from a particular place rather than to go towards anywhere in particular." If they did not "fit harmoniously," these newcomers caused "disastrous results." Zoologists in Europe complained about the arrival of American gray squirrels and other North American species (a "terrific invasion of aliens," as one expert put it in an early 1930s

# Wildlife on the move

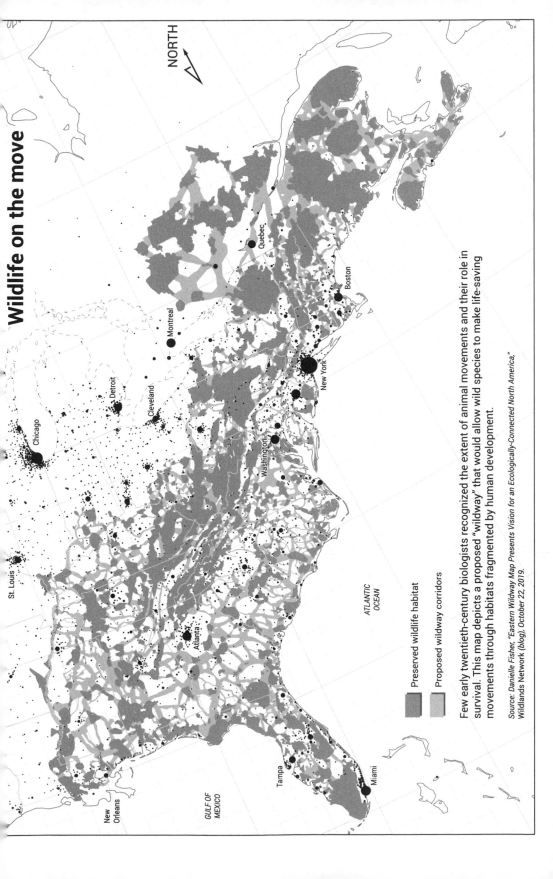

NORTH

St. Louis

Chicago

Detroit

Cleveland

Montreal

Quebec

Boston

New York

Washington

Atlanta

New Orleans

Tampa

Miami

GULF OF MEXICO

ATLANTIC OCEAN

Preserved wildlife habitat

Proposed wildway corridors

Few early twentieth-century biologists recognized the extent of animal movements and their role in survival. This map depicts a proposed "wildway" that would allow wild species to make life-saving movements through habitats fragmented by human development.

*Source: Danielle Fisher, "Eastern Wildway Map Presents Vision for an Ecologically-Connected North America," Wildlands Network (blog), October 22, 2019.*

BBC radio series); in the United States they complained about the intrusion of English sparrows and starlings. (A "European invasion of America is upon us," one zoologist wrote in the *New York Times*. The starlings were "bad citizens," and "undesirable aliens," according to government officials.) The ecologist Aldo Leopold, a friend of Elton's, railed against the "thoughtless importation of Mexican quail" and how they "diluted the hardy northern bobwhite blood in Massachusetts," as a journal article entitled "Game System Deplored as a Melting Pot" reported.

In Germany, people purged plants deemed foreign from the landscape. Nazi leaders instructed locals to banish "foreign" plants from their gardens and to practice a new kind of landscape design in keeping with their superior race. (The traditional gardens many kept, the Nazi garden architect Willy Lange lamented, were characteristic of the inferior "south Alpine" race.) Heinrich Himmler, in addition to masterminding the genocide of millions, issued rules for landscape design forbidding the use of any plants deemed "nonnative." The head of the Reich Central Office for "Vegetation Mapping" called the delicate flowering herb *Impatiens parviflora* a "Mongolian invader" and recommended its extermination. The Nazis zealously protected wild species considered "native." Under their regime, killing an eagle was a crime punishable by death.

Elton did not explicitly extend the implications of his ideas about the dangers of wild migrants to human migrants. But he generally considered his findings about population movements and cycles to elucidate universal principles applicable beyond the specific wild species in which he'd discovered them. For Elton, the line separating nature and human society, as the historian Thomas Robertson put it, "was often a thin one." He made that clear in the way he described animal behaviors in terms generally reserved for humans. At one point, for example, he referred to the lemming migration as a "rather tragic procession of refugees, with all the obsessed behavior of an unwanted stranger in a populous land."

Elton's ideas shed "considerable light on the way the human population should be regulated," one well-heeled Elton fan sniffed. Principles such as Gause's Law "ha[ve] applications in many academic fields of study," added the University of California ecologist Garrett Hardin. Accepting its premises would bring about "a renaissance of understanding."

By the 1930s, the popularity of eugenics had started to diminish in the United States, even as it gained momentum in Germany and elsewhere in Europe. Anxieties about newcomers subsided with the closing of the borders, and the Depression dulled enthusiasm for talk of superior races and their automatically superior lives.

But scientists did not abandon their suspicions about the abnormality of migration in the closed-border world that Linnaeus had described. The New York Zoological Society provided Elton with funding for his research at Oxford. Madison Grant had ascended to the presidency of the society in 1925.

∾

Scientific portrayals of wild migrants as sacrificial zombies and malevolent intruders succeeded, in large part, because the true scale and extent of the journeys migrants undertook remained obscure.

Purposeful, dynamic movements swirled all around, both slow and steady and grand and dramatic. Tiny monarch butterflies weighing no more than half a gram fly three thousand kilometers between eastern North American and central Mexico, where they meet en masse on stands of fir trees. Bar-headed geese soar over the jagged peaks of the Himalayas, climbing from sea level to six thousand meters at a rate of over a kilometer an hour, in the thin, cold night air. Eels in the Sargasso Sea metamorphose into unrecognizable shapes and colors in preparation for grand journeys across the Atlantic.

But scientists can't study migratory behavior in the lab. A rat's ability to learn can be demonstrated by trapping it in a box in which it has to pull a lever to get a morsel of food. A monkey's need for maternal affection can be shown by enclosing it in a wire cage with a bottle of milk and a terry-cloth-covered sculpture. But a creature's drive to migrate can't be so easily re-created in a box or cage. And scientists had few techniques by which to observe it in the wild.

Many of the billions of birds that migrate across seas and continents every year fly under cover of darkness. They can be glimpsed by observers only when in the right place at the right time. At Point Pelee in Ontario, where a narrow spit of marshland extends into Lake Erie, millions of monarchs can be briefly sighted flying south. Along the coasts of the Panama Canal, over half a million raptors can be seen migrating overhead. Along the five-kilometer-long reef in Falsterbo, Sweden, observers can catch nearly 2 million migrating birds from twenty-five different species aloft. In certain secret places, if one looks up into the night sky, tens of millions of migrant birds can be seen flying across the face of the moon; in a single night, 50 million migrants may pass overhead, traveling in a two-hundred-kilometer front.

Most of these spectacles, in which the peculiarities of geography force migrants to briefly merge into dense concentrations, remain as hidden as a tropical beach on the far side of a cave. Given the conventional worldview of an immobile world, few even thought to look.

That started to change when British engineers figured out how to send radio waves into the atmosphere and analyze the echoes they made on passing objects. By installing the technology—nicknamed "radar" for "radio detection and ranging"—they could track all manner of once-hidden movements. During the Second World War, British radar stations installed up and down the coast tracked the movements of enemy planes and ships.

One evening in March 1941, as bombs rained down on London, a radar operator picked up a massive formation of flying objects moving slowly across the English Channel. Presuming the formation indicated an onslaught of invading Germans, military authorities put the Royal Air Force on red alert. The blips continued to approach. When they arrived within forty miles of the Dorset coast, British pilots climbed into their cockpits and flew out over the darkened channel, under orders to intercept and shoot down the intruders.

But when the pilots arrived at the spot from which the radar signals originated, the only sound above the rippling waters was that of their own engines. The night sky was clear, with not an enemy plane in sight.

The confused pilots returned to their base to learn that the blips had mysteriously splintered into single echoes and then faded away.

As the war progressed, the strange signals continued to plague radar installations. Military units would be sent into high alert, only for the signals to inexplicably expand into rings and concentric rings and slowly vanish into nothingness. Whatever created the signals "defied all the known laws of aerodynamics," the *New York Times* reported. The signals arrived day and night. They moved against the wind. Sometimes they moved even faster than the wind.

The ornithologist David Lack had a theory about the strange signals. Lack and Elton both had studied under Julian Huxley and had ascended to the top of neighboring research institutes at Oxford. But the two scientists worked and lived alongside each other like two parallel lines, never intersecting. While Lack was a devout Christian and a music-loving bird-watcher who puttered around in a poorly maintained moss-covered car, Elton zoomed around on motorcycles and in airplanes. Their respective research institutes were situated next door to each other, and their personal residences were separated by just one hundred yards, but despite their professional and geographic proximity, the Lacks and the Eltons never socialized. At Oxford, they always kept the door between their adjoining institutes locked. Their ideas about migration and its role in nature would be similarly distant.

Like other scientists, Lack had been drafted into the war effort, serving in a special unit working on radar technology. He brought his ornithological expertise to bear on the problems he encountered. He'd been watching birds for years. As a student, he'd sit on his bed playing guitar and eating boiled eggs whole, with the shell, for the extra calcium. As a research scientist at Oxford, he spent much of his time in a long, shabby raincoat, hiding among trees to furtively observe his favorite creatures. Thanks to his long hours of bird observation, he knew that birds in groups could fly at speeds similar to those of a fast-moving ship. Even a flock of starlings could cause echoes that might confuse radar operators; large seabirds and shorebirds such as gulls and gray geese could, too.

Lack was not afraid of upsetting the conventional wisdom in order to stick up for his scientific findings. Once he'd suffered the wrath of an elderly bird-watcher who insisted that the same robin had been living in her back garden for seventeen years. When he explained that that couldn't be true because robins didn't live that long, she beat him on the head with her umbrella.

But military officials scoffed at Lack's suggestion that the strange radar signals could be caused by flying birds. Like most people, they believed that birds couldn't fly much at night, because they'd collide into trees and other objects. Plus, birds were small and delicate, nothing like the German fighter jets with 1,200-horsepower Daimler-Benz engines that radars tracked. How could such tiny and insignificant creatures compare to the miracles of wartime engineering?

The spooky signals, they decided, must be ghostly echoes from fallen soldiers, temporarily revived from the beyond. They called them "radar angels."

Years passed before Lack confirmed that the radar angels were, in fact, birds on the wing. One night he and his colleagues rushed out to investigate some radar angel signals and discovered a stand of trees covered in starlings. As they watched, the birds suddenly rose as one and landed on another stand of trees, in a concentric circle around the first—exactly as the unexplained radar signals had recorded.

Lack went on to revolutionize the study of birds, using radar technology to reveal their long-hidden movements. But some-times all it took to catch a glimpse of the migratory world all around was a willingness to look.

One afternoon after the war ended, Lack and his wife set off for a hike into one of the rare high passes through the Pyrenees, which separate France and Spain. They did not expect to see many flying migrants. Creatures such as songbirds avoid mountain passes, and weak flyers like butterflies can hardly withstand the winds whipping around the peaks.

It took four hours for the couple to reach 2,300 meters. Some years earlier Lack had been struck with Bell's palsy, paralyzing one side of his face. Critics described his face as bent into a permanent sneer, but it undoubtedly transformed into something else that afternoon. As the two rested at the high pass, they saw a mass moving toward them in the sky. As they watched in shock and delight, thousands of butterflies and hundreds of songbirds rushed by.

They were flying against the wind and over the mountains. Even Lack, with all his knowledge of and appreciation for migrants and migration, had underestimated the physical capacities and drive of creatures on the move.

～

Before the war, Elton's concerns about the ecological threat posed by migrants had been tempered by his sense that most species stayed put. But as armies of soldiers crisscrossed Europe, aided by new transportation technologies, he started to suspect otherwise.

During the war, he'd been enlisted to help protect Britain's dwindling food supply from rodents. Wartime propaganda cast such rodents as "practically in league with the Nazis," as a modern critic put it. As the creatures he once studied became the subject of extermination campaigns, his assessment of the threat posed by animals on the move took on a new tenor. Everywhere he looked, he saw wild creatures migrating into new places, and precipitating disaster.

In the United States, he noted, the arrival of Asian chestnut trees had introduced a parasitic fungus called *Endothia parasitica*, causing "chestnut blight," which nearly wiped out eastern American chestnut trees. In Europe, a landowner in Czechoslovakia had introduced five North American muskrats. They grew into a population of millions, rampaging through croplands and burrowing into the banks of rivers and streams. In the midwestern United States, the construction of canals introduced the blood-sucking sea lamprey into the Great Lakes. The local lake trout population collapsed. Elton called it "one of the great historical convulsions in the world's fauna and flora."

In his postwar books, radio addresses, and papers, Elton enlisted the language of war to sound the alarm. "It is not just nuclear bombs and war that threaten us," he proclaimed. Wild migrant "invasions" caused "explosive violence." They launched "surprise attacks," with "attack[s] and counterattack[s]." And their goal was the same as that of the Nazi invaders his fellow Britons had fought during the war: complete domination or, as he put it, "the eventual expansion and occupation of territory from which they are unlikely to be ousted again."

Even if newly arrived species seemed benign, he said, their arrival presaged danger. They might just be lying in wait, ready to "rapidly spread to become pests of major proportion." In time, the invaders would take over, displace original inhabitants, and leave the entire ecosystem diminished. Wild invaders, he warned, would "eventually reduce the rich continental faunas to a zoned world fauna consisting of the toughest species." They'd trigger a "zoological catastrophe."

To depict species on the move as "invaders" and their impact as catastrophic, Elton cherry-picked only the most disruptive of introduced species. He also considered only the costs that introduced species exacted, and none of the benefits they rendered, of which there were many obvious ones—the harvests from corn, soybeans, wheat, and cotton, to name just a few, all of which derived from plants that had been introduced from one continent into others.

As GPS technology would later allow biologists to document, wild species, like peoples, move around all the time, whether through their own locomotion or by being carried on the winds, currents, or backs of other moving creatures. Inserting themselves into continuously shifting landscapes, most either fail, struggle, or incorporate themselves inconspicuously in temporary assemblages that themselves assort and reassort dynamically. (Which is why biologists would also find that the introduction of a species from one landscape into another generally increases biodiversity.) It is true that predators and pathogens introduced into relatively closed ecosystems, such as lakes and islands, can drive already-resident species to extinction. But most ecosystems do not have closed borders around them.

Elton delivered his warnings about invasive species in a series of BBC broadcasts, straightforwardly titled *The Invaders*. Then he wrote a short book on the subject. The book was "written hurriedly," writes the historian Matthew Chew, and "less coherently and deeply considered" than his other works. Elton knew it, admitting to one of his students that he'd pounded out the book on the basis of the radio addresses in a matter of weeks. "I did the broadcasts," he wrote, "am now turning that stuff into a 45,000-word heavily illustrated book . . . which I shall have written in nine weeks."

His 1958 book, *The Ecology of Invasions by Animals and Plants*, would inform the management of national parks and programs protecting wildlife around the world. It launched a whole new field of inquiry dedicated to documenting the negative impact of species on the move, a field known as "invasion biology," which would take off in the 1980s. The book would be hailed for decades to come as "one of the central scientific books of our century," as the science writer David Quammen called it in 2000.

∽

The spectacle of the suicidal migration of the lemmings embedded itself into public consciousness in 1958, the same year *The Ecology of Invasions* came out.

Produced by Walt Disney studios, the nation's most powerful and celebrated production company, *White Wilderness* showcased the eerie, rarely seen world of the frozen Arctic, using the most cutting-edge cinematic techniques of the day. Nine photographers had "roamed the snow-sheeted wastes," covering "thousands of miles of tundra, lakeland, mountains and icy rivers" to make the film, the *New York Times* reported. The documentary, one of the most anticipated of the year, was the thirteenth in a series show-casing the strange and mysterious phenomena of the natural world, with stunning, never-before-seen footage from some of the most remote places on earth, from the islands of the Bering Sea to the plains of the Serengeti. In *White Wilderness*, audience members would be treated to "one of the most ambitious sagas in the annals of nature photography," one of the film's consultants told the *Times*.

Filmgoers settled into the Normandie Theatre's plush, mohair-velvet-upholstered seats one stormy August afternoon in New York to watch the premiere. They'd be talking about the documentary's most shocking images for years to come. The six-minute scene depicted one of "nature's strangest phenomena," as Disney put it.

In it, the camera pans across an icy landscape where lemmings congregate. With their long whiskers and lush fur, they look like round, fluffy hamsters. At first they meander, delicately sniffing the ground and each other. But then they slowly start to move across the frozen tundra, their tiny paws flicking bits of snow behind them as they gain momentum.

The narrator explains that a cliff—not pictured in the scene—lies in their path. "Ahead lies the Arctic shore and beyond, the sea," the baritone voice-over says, as the furry horde continues its determined march. "And still the little animals surge forward."

The camera follows along. "They reach the final precipice," the narrator says. "This is the last chance to turn back."

Few in the audience could boast any special expertise in lemming behavior. Still, every schoolchild knew that the guiding principle of animal behavior, as for human behavior, is self-preservation. On screen, the lemmings reach the brink of the rocky cliff, their sharp claws gripping the edge. They pause.

One by one, they leap.

The next image is of furry balls aloft, little feet scrambling in the air.

The camera then moves to the gray, featureless expanse of the sea below the cliff, as it's punctured by a rapid fusillade of falling lemming bodies, plunging into the icy depths to their deaths.

It's a ghastly scene. Did they lemmings fall by mistake? Was their plunge the result of some kind of bizarre navigational error? viewers likely wondered. No. The lemmings' "eerie death march" stems from a "blind, instinctive impulse," the filmmakers explained. Their mass suicide is no anomaly or accident. Flinging themselves from cliffs is what the lemmings are supposed to do. "It is not given to man," the narrator says, "to understand all of nature's mysteries."

Elton is remembered today as the "founding father" of animal ecology and a "towering figure" in biology, as the Royal Society put it. But though he circled back to his story of lemmings again and again, neither he nor any of his colleagues had ever seen a lemming migration into the sea. After decades of attempts to observe the phenomenon, scientists could boast just a single blurry photograph, picturing a lemming in transit. In 1935 one of Elton's students, Dennis Chitty, ventured into the Canadian Arctic with plans to study lemming migrations, expecting to find the land "overrun with lemmings" as he put it. He traveled aboard a ship up the Hudson Bay, around Baffin Island, north to Ellesmere Island, then west to the Northwest Passage. For seven weeks, they sailed through the blustery cold, covering thousands of miles. They saw "masses of ancient lemming droppings" but not a single lemming. Fifty years later, Chitty tried again, in Finse, Norway. He'd heard reports that lemmings had boomed that year. By the time he arrived, they'd vanished.

The truth about lemmings emerged when biologists peered under the snow in lemming territory. It turned out that they didn't disappear into the icy depths of the Arctic Sea. They dug holes and hid under the snow, feeding on moss and breeding their young in the little gap created by the warm ground melting the layer of snow directly above it known as the "subnivean space." For years, biologists hadn't thought to look, because breeding under the snow had been considered biologically impossible.

During snowy years, unknown to anyone scanning the landscape, their numbers build up under the snow cover. When the snow melts, their tunnels and holes fill with meltwater, forcing them to leave abruptly. Suddenly, they appear in great numbers as if out of nowhere and their tracks speckle the sea ice. Then as winter approaches and snow blankets the land, they burrow back under the snow, seeming to mysteriously vanish.

Collett's book, which Elton had relied on, was like Linnaeus's source material, an amalgam of myths, legends, and "cock-and-bull stories from Norwegian sailors," the historian of ecology Peder Anker explains. Elton might have realized it if he hadn't relied on his crude translation of the book.

But while scientific evidence for the suicidal lemming migration had materialized out of a series of misunderstandings, its popularization was the result of a purposeful deception. In 1982, the Canadian Broadcasting Corporation aired a documentary called *Cruel Camera*, about animal abuse in film. It recounted how the lemming migration in *White Wilderness* had been filmed.

The lemming suicide march had been staged. The Disney filmmakers used an animal trainer who built little studios and sets for their wildlife documentaries. He had created scenes of geese flying, for example, by positioning captive geese in front of wind machines. The animal trainer and his team hired local kids to capture lemmings for twenty-five cents per lemming. Then they shipped the lemmings one thousand miles away to a set they'd built outside Calgary, replete with a painted Arctic sky. They herded the lemmings onto a turntable, then filmed them running around it. This created the appearance of a horde of lemmings running in a straight line.

Then they gathered the lemmings, loaded them onto a truck, and with the cameras rolling, tipped them out over a riverbank. The lemmings hadn't committed suicide. They'd been murdered.

∼

*White Wilderness* infiltrated the public mind for decades before the exposé came out. The scene of the lemming suicide helped make *White Wilderness* a critically acclaimed hit. In 1959 the film won the Academy Award for best documentary feature. It would be shown in public schools across the country for years, bringing to the masses Elton's bleak vision of the ecological necessity of migrant deaths.

I learned about the lemmings' mass suicide in a fluorescent-lit classroom of a suburban Connecticut middle school in the late 1970s. Like everyone else, I found the story darkly fascinating. I remember being struck by how much the driven lemmings looked like my own pet hamster, Hammy, whose reputation as a simple-minded creature driven by immediate needs seemed suddenly and thrillingly suspect.

The lemmings' macabre migration captivated the nation. Their mass suicide quickly became a cultural shorthand for self-destructive behavior of all kinds. A musical group from Berkeley, California, dubbed themselves The Lemmings and published an image of a line of cars driving off a cliff into the sea on the cover of their album. The *New Yorker* cartoonist James Thurber imagined a conversation between a lemming and a scientist in his "Interview with a Lemming." "I don't understand," says the scientist, "why you lemmings all rush down to the sea and drown yourselves." "How curious," says the lemming. "The one thing I don't understand is why you human beings don't." The English poet Patricia Beer waxed darkly about the lemmings' "hot blood" pouring into the "cold sea." During the war, millions had marched to their deaths "like lemmings," as the psychologist Bruno Bettelheim put it. "War," the biologist Richard A. Watson wrote, "is the ultimate manifestation of the lemming-like madness that grips large populations of men in great need."

In the late 1950s, when the film first came out, the notion that suicidal migrations occurred in nature helped make sense of the still-raw trauma of the Second World War. By sacrificing themselves, the soldiers and others who died had helped maintain a balance in nature just as the migrant lemmings did by flinging themselves into the sea. The appropriate conclusion to the migratory act, in other words, was death.

With the borders around countries such as the United States firmly closed, no one had cause to ponder the political and ecological dilemmas that might arise if migrants didn't martyr themselves.

That would change.

# 6

# MALTHUS'S HIDEOUS BLASPHEMY

B etween the 1940s and '60s, scientists documented another knotty ecological phenomenon in nature, one that leading population biologists would call on to warn about the dangers posed by migration.

Aldo Leopold, a friend of Elton's, wrote the first reports about Kaibab in 1943. He called the episode the result of an "upset" in the "balance of natural forces in the ecosystem."

The Kaibab, an isolated thousand-square mile high-elevation plateau bounded by deep canyons in northern Arizona, had been established as a game preserve in 1906. To expand the populations of deer that hunters liked to stalk, the U.S. Forest Service endeavored to clear the Kaibab of the deer's predators. Between 1907 and 1923, it killed 3,000 coyotes, 674 mountain lions, 120 bobcats, and 11 wolves.

The cull transformed the plateau. Liberated from the appetites of their predators, the deer population skyrocketed. At the beginning of the twentieth century, about four thousand deer lived in the Kaibab. By 1924, observers estimated that the deer population had grown to one hundred thousand.

But the deer population did not thrive for long. Their success planted the seeds of their demise. The deer stripped the aspen, spruce, and fir of their bark, stunting the trees' growth and diminishing the quality of vegetation available for them to eat. They began to starve. "In nearly every case," observers reported in 1924, "the outline of the ribs could be easily seen through the skin." Between 1924 and 1928, nearly three-quarters of the population's fawns perished.

A similar upset occurred on St. Matthew Island, a formidable slab of thousand-foot cliffs surrounded by the Arctic seas of the Bering Strait. During the war, the Coast Guard had captured twenty-nine reindeer from hundreds of miles away, floated them on a barge, and deposited them on the tiny narrow island, to serve as a backup food supply for soldiers manning the small radio navigation station they'd briefly set up there. After the war, the station was dismantled and the reindeer, who'd escaped the appetites of the radio navigation staffers, were left to their own devices on the lichen-rich, predator-free island.

Researchers arrived to check on them in 1963. They found the island crisscrossed by reindeer tracks and droppings. The initial population had grown to more than six thousand. The reindeer had trampled the lichen that sustained them nearly beyond recovery. A few years later when the researchers visited again, they found little sign of the reindeer. All that was left was their bleached skeletons.

The deer of Kaibab and St. Matthew did not sacrifice their surplus numbers by migrating off cliffs to their deaths. They did not ritually murder every other baby like the Funafuti or engage in battles that killed scores.

They just kept consuming and reproducing. If allowed to freely roam, they'd bring their outsized, ecosystem-killing appetites elsewhere.

∽

What happened on Kaibab and St. Matthew recalled the warnings of the eighteenth-century cleric Thomas Robert Malthus. Malthus had pointed out that aiding the poor with food and clothing, as England's Poor Laws required and which most people viewed as beneficial charity, circumvented nature's checks on human population growth. "We cannot, in the nature of things, assist the poor, in any way," Malthus wrote in 1798, "without enabling them to rear up to manhood a greater number of their children." Progress against poverty, disease, and hunger would result in an exponentially growing human population, Malthus warned, whose appetites would regularly outstrip the food supply, creating a permanent state of conflict and scarcity.

Malthus was proved wrong in the centuries that followed his warning. All the things that caused death rates to fall—modernization, economic development, prosperity—also led to declining birth rates. Social scientists call the shift from high birth and death rates to low ones the "demographic transition." In the United States, for example, modern sanitation and other factors drove death rates down from 25 per thousand in the seventeenth century to less than 10 per thousand a few centuries later. But catastrophic population growth was averted because birth rates fell, too: the average number of children born to white American women declined from seven in 1800 to just two by 1940. Influential thinkers condemned Malthus as an alarmist. The nineteenth-century philosopher Friedrich Engels called his theory "hideous blasphemy against nature and mankind."

But then, after World War II, demographic trends shifted. In the United States and other prosperous countries, birth rates zoomed. The "baby boom" reversed the prediction of demographic transition theory, which suggested that people in prosperous societies tended to have smaller families. Meanwhile, in poor countries such as India, death rates fell. Chemicals such as fertilizers and antibiotics, developed during the war, circumvented the disease and famine that otherwise killed off millions. That undermined demographic transition theory, too, because there'd been no underlying economic development or modernization.

Scientists started resurrecting Malthusian concerns in a spate of popular books. The result for humans could be a slow-motion version of the collapse on the tablelands of Kaibab, they warned. The ornithologist William Vogt, who'd written the foreword to John James Audubon's illustrated classic *Birds of America*, wrote a best-selling book on the topic called *The Road to Survival* in 1948. A book exploring similar themes written by Henry Fairfield Osborn's son, Henry Fairfield Osborn Jr., who'd taken over the helm of the New York Zoological Society, appeared that same year.

"Every argument, every concept, every recommendation" in Vogt's book, notes the historian Allan Chase, became "integral to the conventional wisdom of the post-Hiroshima generation of educated Americans." The *Reader's Digest*, a publication with sales second only to the Bible, reprinted a condensed version of Vogt's book. Between 1956 and 1973, seventeen out of twenty-eight general biology textbooks included a version of Leopold's account of the collapse of the deer in Kaibab.

∾

Tall and angular, with piercing eyes framed by heavy brows and long sculpted sideburns, the Stanford University biologist Paul Ehrlich had grown up in New Jersey collecting butterflies and wandering around the American Museum of Natural History.

Like others of his generation, he knew about what had happened on the Kaibab and on St. Matthew Island. He had absorbed the lessons of Vogt's and Osborn's books, which he'd read while an undergraduate at University of Pennsylvania. Vogt had even given a lecture on campus while he was a student.

But Ehrlich studied checkerspot butterflies. They were nothing like the deer of the Kaibab. Female checkerspots laid hundreds of eggs, but in many years getting even two to survive was a stretch. If it was a little too warm, or a little too rainy, the larvae's growth would fail to coincide with the brief blossoming of their food supply. The plantago plants the larvae fed on were not only short-lived, they also grew exclusively in outcrops of serpentine rock in the arid grasslands of the hills, a unique and scarce habitat. For a population of checkerspots to survive, a newly emerged butterfly not only had to find those special plants and lay her eggs on them, she also had to do it at the right time and under the right conditions, so that the eggs would mature into larvae and be able to feed on the plants before the plants died down for the winter. What biologists called the "phenological window"—the time between the emergence of an adult butterfly ready to mate and the death of the plants her offspring can feed on—could be measured in a matter of days.

When conditions worsened for checkerspots, their populations simply died down and vanished. Despite the wings that allow them to surmount geographic and other barriers and spread into new regions, checkerspot butterflies, Ehrlich knew, were isolated homebodies, rarely leaving their patch of mountainside, regardless of how bad conditions got. He'd conducted studies that proved it, noting what he called their "remarkable lack of wanderlust."

So long as his attention remained focused on butterflies, the ecological calamities of population growth predicted by Malthus—famine, environmental degradation, and collapse—did not concern Ehrlich. Then he visited South Asia.

～

Ehrlich landed at dusty Palam Airport in New Delhi, India, in the deep heat of the pre-monsoon season in late June 1966. It was the last stop on a year-long multicountry research trip on which he'd brought his wife, Anne, and daughter, Lisa Marie, along.

Most American visitors found the place, while dizzyingly hot and turbulent, charming and transformative. Shaggy-haired counterculture enthusiasts flocked to India to imbibe its ancient traditions of yoga, meditation, Buddhism, and more. The members of the Beatles arrived in Delhi about a week after the Ehrlichs to pick up authentic Indian classical musical instruments.

Not Ehrlich. Everywhere he looked, he saw a disaster in the making.

"The streets seemed alive with people," he'd write later. "People eating, people washing, people sleeping. People visiting, arguing, and screaming. People thrusting their hands through the taxi window, begging. People defecating and urinating. People clinging to buses. People herding animals. People, people, people, people." Delhi dramatically conveyed "the feel of overpopulation," he wrote, calling the city "hellish."

Repulsed, the Ehrlichs left Delhi and headed north to the thickly forested high-altitude valleys of Kashmir, nestled between towering Himalayas. Here they encountered yet more signs of impending ecological catastrophe. Kashmir's high-altitude meadows had become "biologically barren," Ehrlich wrote, "grazed to within a fraction of an inch above the ground." He could find hardly any of the butterflies he'd hoped to study. Plus, the hotel the family stayed in was dirty; the houseboat they rented on Kashmir's famous Dal Lake was grossly overpriced. Kashmir was a "big disappointment," Ehrlich wrote to a friend.

For Ehrlich, the Indians had effectively trampled on their own lichen, like the reindeer of St. Matthew. If it hadn't been for a shipment of 9 million tons of wheat from the United States the year before his visit, Ehrlich figured, India would surely have been plunged into famine already, just as the deer on Kaibab had.

In fact, the crowds and environmental damage the Ehrlichs saw had as much to do with local economic and political factors as with the growth rate of the population. India's population had grown, but the city of Delhi had not, in fact, become especially large compared to other cities around the world. With a population of 2.8 million, Delhi was just a fraction of the size of, say, Paris, where 8 million resided. The chaos and crowding he saw resulted less from the size of the local population than from the effects of new government programs that encouraged people from the rural hinterlands to move into the city for factory jobs. The influx of newcomers had overwhelmed local housing capacity and the city's rudimentary infrastructure.

And while Kashmir had indeed suffered environmental damage, attributing that to population growth was a stretch, too. The Ehrlichs had arrived less than a year after India and Pakistan had fought a bloody and destructive war over control of the region. Tens of thousands of troops armed with artillery and tanks had overrun Kashmir's steep valleys, turning vast stretches into battlegrounds. So many soldiers had poached the local wildlife for food and for fur that many of the valley's unique species had been pushed to the verge of extinction. Kashmir may have been "spoiled," as Ehrlich pointed out, but destructive mountain warfare had been a powerful culprit in the process.

For Ehrlich, such historical particulars obscured the forest for the trees. In close-up, conditions in India may not have conformed to Malthus's predictions, but through a wide-angle lens they did. The population grew; the quality of the environment declined. It was that simple, and it was just as Malthus and Vogt had said.

∾

Ehrlich's crusade to arrest the growth of the human population began soon after he arrived back home from India. At first, he delivered his warnings about the risks of Kaibab-like collapse in humans to students at Stanford. Soon local clubs and NGOs started inviting him to speak to their members.

Ehrlich's Stanford offices became a hub for scientific debate over the ecological crisis precipitated by population growth. At weekly seminars and conferences at Stanford, he gathered scientists such as the University of California ecologist Garrett Hardin, the social scientist Kingsley Davis, and others to share notes on the possibility of a human population explosion and its portents for the future.

They saw signs of impending Kaibab-like collapse all around, including in California. By 1962, more people lived in California than the state of New York. Red brake lights glared on the jammed freeways. The cities sprawled. And when the unprocessed tailpipe emissions of millions of California cars streamed into the Golden State sunshine, a toxic chemical reaction created clouds of smog, which because of their location in mountain-ringed basins, hung thickly over California's major cities. The air grew so polluted it became opaque. People walking the streets of Los Angeles took to wearing "smogoggles" to protect their eyes.

Pollution wasn't the only ominous sign. New research suggested that, like famine and environmental degradation, antisocial behavior might be a sign of impending Malthusian collapse, too.

In his influential study, the Johns Hopkins animal behavior expert John B. Calhoun built a quarter-acre enclosure behind a neighbor's house outside Baltimore, Maryland, into which he'd released five pregnant rats. With no predators, the rat population grew into the hundreds. But Calhoun continually replenished their food supply. He would not let the rats eat themselves into Malthusian collapse.

Chaos ensued regardless. The rats' behavior changed. Male rats banded together in aggressive groups and attacked females and young rats. They ate the bodies of the dead. Female rats neglected and even attacked their own infants. Some rats became homosexual; others hypersexual. Eventually the rats' behaviors became so disturbed that they could no longer successfully reproduce, presaging the population's ultimate collapse.

Ehrlich encouraged a social psychologist named Jonathan Freedman to conduct follow-up studies to see if the effect occurred in humans. Even before Freedman published his results, Ehrlich cited what he considered their foregone conclusions in his own papers. He wasn't alone. In Congress, politicians sounding alarms about human population growth offered Calhoun's work to buttress their arguments, too. Commentators used Calhoun's results to link antisocial behavior with crowds. "Uncontrollable aggressiveness" resulted from crowding, the zoologist and TV host Desmond Morris wrote. It had been "proved conclusively with laboratory experiments." The journalist Tom Wolfe compared crowds of New Yorkers "running around, dodging, blinking their eyes, making a sound like a pen full of starlings or rats." The critic and philosopher Lewis Mumford wrote of the "ugly barbarization" of humans due to "sheer physical congestion," which had been "partly confirmed" by Calhoun's experiments in rats. The "freedom to breed," Hardin argued, had become "intolerable."

Scientists started to refer to human population growth not as the happy outcome of prosperity and improved health but as a silent killer that would violently erupt. *Science* magazine called population growth the "P-bomb," like the A-bomb and the H-bomb. *Time* magazine featured the problem in a 1960 cover story titled "That Population Explosion."

At the time, frightened American readers could take comfort in the presentation of the problem as a distant one, transpiring a world away. *Time* magazine suggested as much, illustrating the story with a collage of bare-breasted African women and sari-wrapped Indian women laden with infants. Americans were effectively cordoned off from such people: immigration laws had prevented their kind from penetrating U.S. borders for decades.

~

After refusing entry to ships full of terrified Jews and others persecuted by the Nazis, chastened political leaders in the United States, Europe, and elsewhere belatedly agreed to provide refuge to those fleeing Nazi-style persecution, signing the United Nations' Refugee Convention in 1951. With the civil rights movement gaining momentum, pressure built to dismantle the decades-long racial quotas on the borders. Two years after the *New York Times Magazine* excerpted President John F. Kennedy's inspirational essay dubbing the United States a "nation of immigrants," Congress passed the Hart-Celler Act in 1965, removing race as a criterion for judging whether a migrant could enter the United States.

The Hart-Celler Act had been sold as a practical measure, to allow skilled foreigners such as my parents help the country staff newly expanded government programs such as Medicare and Medicaid, and to bolster the nation's reputation as a welcoming nation, in contrast to the Soviet Union's closed society. It had not been intended to actually alter the racial makeup of the nation. Most likely, its architects thought, white-skinned Europeans would continue to dominate the migrant flow, just as they had in the past. This is "not a revolutionary bill," President Johnson said upon signing it, promising that it "will not reshape the structure of our daily lives." "There is no danger whatsoever of an influx from the countries of Asia and Africa," Representative Emanuel Celler reassured the nation.

He was wrong. Nine out of ten of the post-1965 newcomers hailed from Asia, Latin America, and other non-European locales. The explosive chaos described by the Malthusian ecologists emanating from countries like India could reach the shores of the United States. The population bomb would not be contained. Its catastrophic effects would be "compounded," the Ehrlichs wrote, "by man's unprecedented mobility."

A popular movement to defuse the population bomb—and to contain its effects outside American borders—started to coalesce.

∼

The crusading leader of the Sierra Club, David Brower, heard Ehrlich being interviewed about the population problem on a daytime talk show. Inspired, he called the publisher Ian Ballantine. Together they persuaded Ehrlich to write a popular book on the subject.

Paul and his wife, Anne, wrote it together. Anne, a French major in college, had often collaborated with him on his projects. She'd even illustrated his doctoral thesis on insecticide resistance. But when Ballantine published the book, they decided to drop her name from the byline for marketing reasons. They also changed the title from the dryly descriptive *Population, Resources, and Environment*, which the Ehrlichs had come up with, to the catchier *The Population Bomb*.

In it, Ehrlich dispensed with the scientist's normal caution and circumspection. He warned that population growth would usher in an "utter breakdown of the capacity of the planet to support humanity" within fifteen years. By 1984, he predicted, Americans would be dying from dehydration.

Population growth was a problem that implicated everyone, but Ehrlich prescribed dramatically different solutions for Americans than for foreigners. While Americans needed to have their awareness raised about their reproductive habits and consumption patterns, he suggested that all Indian males with three or more children be sterilized—and recommended sending in American helicopters, doctors, vehicles, and surgical instruments to help the Indians do it. He suggested "very unpopular foreign policy positions," such as letting some needy countries starve rather than shipping them food aid and adding antifertility drugs to their water supply.

Ehrlich was not an overt racist. On the contrary, he championed civil rights, and as a scientist, he vociferously objected to the concept of biologically distinct races. He had helped organize an antisegregation protest in Kansas while a postdoc, and he'd written books and papers decrying the psychologist Arthur Jensen and the Nobel Prize–winning physicist William Shockley, who argued that black people were genetically inferior and should therefore be the primary subject of population control efforts. (Unlike his writings on the population problem, Ehrlich's book on race was condemned for an "arrogance and polemicism that rivals the worst excesses of [his] chosen opponents.")

And yet no one who read *The Population Bomb* could miss that Ehrlich didn't consider foreigners to have the same capacity for change and understanding as his own people. Throughout the book, he characterized people's variable practices not as dynamic and responsive to changing conditions but as immutable biological features.

He bemoaned the fact of child marriage in India, noting how it extended childbearing years over decades, for example, but considered the dire policy of forced sterilization more feasible than altering that cultural practice. He considered the difference between castration and sterilization "almost impossible to explain" to people in India. He insisted that the population control achieved in Taiwan, Korea, and Japan could never succeed in poor countries. "We would be foolish in the extreme" to think anything similar could occur in "other parts of Asia, and Africa, or in Latin America," he wrote. He disdained voluntary family planning programs, which left it up to women to decide how many children they wanted. While he himself—an educated Western male—could see the Malthusian problem posed by large families, he could not see how women, especially women who lived in poor countries—or as he sometimes called them, the "never-to-be-developed" nations—ever could.

Ehrlich's characterizations conformed to a popular theory in his scientific discipline. During a brief period in the late 1960s and '70s, a theory called "r/K selection" consumed population biologists. In effect, it imagined two kinds of places—those where the living was easy, and those where the living wasn't—and two broad classes of creatures that inhabited them. In the easy-living places resided "r-strategists," small, quickly maturing, highly fecund creatures, almost mindless in their pursuit of large families. Their environments did not pressure them to be particularly clever or frugal, which is why they were mostly dumb and wasteful. In the tough places resided the large-bodied "K-strategists," whose environments required them to be clever and frugal and who tended to mature later and invest more in their few offspring. Conservation biologists used r/K selection theory to distinguish between r-strategists such as mice, compared to K-strategists such as elephants.

Controversially, in 2000 the Canadian psychologist John Philippe Rushton would explicitly apply r/K selection theory to human racial groups, arguing that black people were r-strategists, "Orientals" K-strategists, and white people in between. Rushton's bias was overt: he served as the head of the eugenics research outfit the Pioneer Fund, which Harry Laughlin had led for years.

Ehrlich's bias was not overt, but his characterizations of Indians and Europeans echoed Rushton's. In his foreword to *The Population Bomb*, the Sierra Club's David Brower similarly presumed fixed r/K–like differences between people who lived in different places. "Countries are divided rather neatly into two groups," Brower wrote. "Those with rapid growth rates, and those with relatively slow growth rates." The fast growers, he wrote, were "not industrialized, tend to have inefficient agriculture, very small gross national products, high illiteracy rates and related problems." The slow growers, presumably, were just the opposite.

Ehrlich's fellow neo–Malthusian scientists, such as Kingsley Davis, explicitly called for an end to immigration. According to Davis, immigrants slowed technological progress and gave rise to "school problems, health risks, welfare burdens, race prejudice, religious conflicts and linguistic differences," as he wrote in *Scientific American*. He recommended closing the state of California to newcomers from Mexico and China.

Ehrlich alluded to the invasiveness of immigration from places like India, too. Starving Indians, he warned, would flood the borders, intent on stealing American resources. "They have seen colored pictures in magazines of the miracles of Western technology," he wrote. "They have seen automobiles and airplanes. Many have seen refrigerators, tractors, and even TV sets. Needless to say, they are not going to be happy." Indians, he wrote, would not "starve gracefully without rocking the boat." Most likely, they'd "attempt to overwhelm us in order to get what they consider to be their fair share."

To prevent them from swamping the nation, muscular action would be required. "I know this all sounds very callous," Ehrlich wrote, sympathizing with his readers while simultaneously priming them to accept the necessity of authoritarian measures. Drugging people and forcing surgeries on them was coercive. But it was "coercion in a good cause."

"Remember the alternative," he wrote.

~

Ehrlich's provocative two-hundred-page book, though witty, dark, and stylishly written, generated modest interest at first. Then in early 1970, Johnny Carson called. Carson's late-night talk fest *The Tonight Show* dominated television, bringing in more money than any other program on television at the time or since. A slot on Carson equaled a shot at stardom, as a generation of performers, including Barbra Streisand, Woody Allen, and Steve Allen, had discovered.

At age thirty-seven, on stage with the biggest name in the entertainment industry, Ehrlich "wasn't particularly nervous," he recalled later. He was a showman who knew how to get people's attention and enjoyed doing so. He'd "always been a loudmouth," he says. Ehrlich's explanations about the Kaibab-like future that awaited and the radical interventions required to avert it were beamed into the homes of nearly 15 million people across the country.

After his first appearance on *The Tonight Show*, viewers sent in five thousand letters, more than any other guest had ever provoked. Sales of *The Population Bomb* exploded. In the first months of 1970, the book sold nearly a million copies; by the end of the year, it had sold nearly 2 million. The journalist Joyce Maynard remembers the "rush of dread" she felt while reading *The Population Bomb*. "Not personal individual fear," she explained, "but end-of-the-world fear, that by the time we were our parents' age we would be sardine-packed and tethered to our gas masks in a skyless cloud of smog."

Ehrlich became a celebrity with the stature of Al Gore or Neil deGrasse Tyson today, overnight. He appeared on Carson's show dozens of times; during one stretch, he appeared three times in as many months, chatting with Carson about a range of political issues facing the nation. "Richard Nixon would do very well on an IQ test," he deadpanned, in a discussion about the uselessness of intelligence tests, "but would you want your daughter to marry him?" "We are quite happily destroying the only resources [our children will] ever have," he'd point out. "But you know," he quipped, "what did posterity ever do for us?"

Prestigious institutions showered Ehrlich with honors—an Emmy nomination, an endowed chair at Stanford, awards from the United Nations, the MacArthur Foundation, and the Royal Swedish Academy of Sciences. Entrepreneurs such as Hugh Moore, who'd made a fortune selling disposable Dixie-brand drinking cups, paid for ads in major newspapers, sent hundreds of college radio stations free radio segments featuring Ehrlich, and distributed hundreds of thousands of leaflets and brochures about *The Population Bomb*.

Top Hollywood directors and actors signed on to shoot sci-fi films inspired by the famished, overcrowded future Ehrlich described. In 1972's *Z.P.G.*, Charlie Chaplin's daughter, the Golden Globe–nominated actress Geraldine Chaplin, portrays a woman who rears a robot baby in a future world in which all human reproduction had been banned for thirty years; in the following year's *Soylent Green*, Charlton Heston navigates a future New York City so depleted and overcrowded that people survive on cannibalistic corporate rations.

Suspicions that population control really meant controlling certain populations and not others dogged the movement from the beginning. At a conference in June 1970, a group of African American activists stormed out, accusing the assembled environmental and population control activists—Ehrlich, Hardin, and others—of being interested less in the problem of population than in problem populations. The conference agenda—which included visiting "teeming urban areas" where "a dangerous crisis of ecological imbalance already exists," as the organizers put it—aimed at the "systematic reduction of a specific population, namely blacks, other non-whites, the American poor and certain non-white and ethnic immigrants," the protesters wrote in a statement.

Their critiques did not slow the population control movement's momentum. For counterculture activists, the movement offered an opportunity to rile religious conservatives by touting the birth control they considered sacrilege. For business leaders who'd built their empires promoting consumerism, it offered a handy diversion from their own role in despoiling the environment. For Western NGOs and their allies in the political establishment, it offered an opportunity to promote technological quick fixes such as contraceptive devices that compared favorably to slow, Soviet-style economic development. Impoverished, rapidly growing countries such as India went along with the new population control regime, too. Their elites didn't mind blaming their country's poverty and hunger on the fertility of poor women, rather than, say, the oppressive caste system or widespread corruption from which they benefitted.

On one of his appearances on *The Tonight Show*, Ehrlich announced the founding of a new organization called Zero Population Growth, aimed at averting Malthusian collapse by making abortion and sterilization more accessible. The group quickly enlisted sixty thousand members. On college campuses across the country, student activists held events in which they tossed condoms into crowds. They conducted "experiment[s] in overpopulation," in which participants were herded together in circumscribed spaces and fed a "famine diet" of rice and tea. They affixed pins on their lapels featuring the symbol for a male with a

# Border walls

This map depicts the spate of border walls built across the globe since the end of World War II. The antimigrant ideology that propelled the construction of Fortress Europe and Fortress America grew in part out of postwar fears of population growth articulated by population biologists.

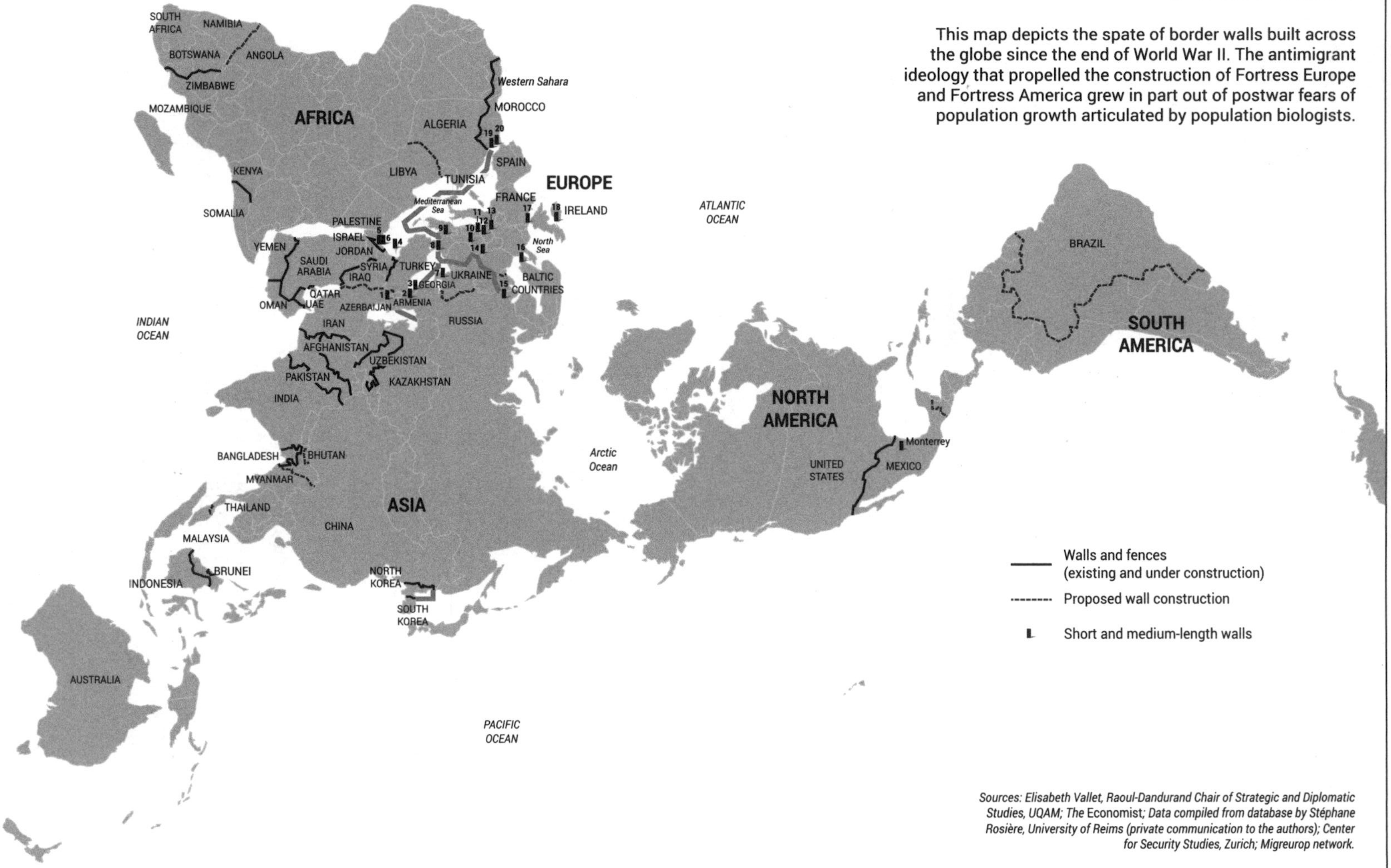

SOUTH AFRICA
NAMIBIA
BOTSWANA
ANGOLA
ZIMBABWE
MOZAMBIQUE
AFRICA
Western Sahara
MOROCCO
ALGERIA
SPAIN
LIBYA
TUNISIA
EUROPE
KENYA
SOMALIA
FRANCE
IRELAND
Mediterranean Sea
North Sea
PALESTINE
ISRAEL
JORDAN
YEMEN
SAUDI ARABIA
SYRIA
IRAQ
TURKEY
UKRAINE
GEORGIA
ARMENIA
AZERBAIJAN
BALTIC COUNTRIES
QATAR
UAE
OMAN
IRAN
RUSSIA
AFGHANISTAN
UZBEKISTAN
KAZAKHSTAN
PAKISTAN
INDIA
INDIAN OCEAN
ATLANTIC OCEAN
BRAZIL
SOUTH AMERICA
NORTH AMERICA
UNITED STATES
MEXICO
Monterrey
Arctic Ocean
BANGLADESH
BHUTAN
MYANMAR
THAILAND
ASIA
CHINA
MALAYSIA
BRUNEI
INDONESIA
NORTH KOREA
SOUTH KOREA
AUSTRALIA
PACIFIC OCEAN
Walls and fences (existing and under construction)
Proposed wall construction
Short and medium-length walls


*Sources: Elisabeth Vallet, Raoul-Dandurand Chair of Strategic and Diplomatic Studies, UQAM; The Economist; Data compiled from database by Stéphane Rosière, University of Reims (private communication to the authors); Center for Security Studies, Zurich; Migreurop network.*

chunk cut out of the circle to broadcast their vasectomies. They wore IUDs fashioned into earrings.

Scientists such as Vogt, Osborn, and Kingsley Davis launched international NGOs such as the Population Council, which shipped 1 million IUDs to India. Philanthropic groups such as the Rockefeller and Ford foundations pressured the U.S. government to demand that foreign-aid recipients crack down on the fertility of poor women as a precondition for receiving funds or food.

In June 1975 the population control movement scored a major victory. That summer the Indian prime minister suspended the constitution and embarked on an ambitious plan to sterilize its booming populace. In states across India, men with more than three living children would be required to undergo sterilization, and pregnant women with three children would have to abort their babies. Government employees working under quotas fanned out across the country, scalpels and IUDs in hand.

~

John Tanton, an unassuming man with hooded gray eyes and an unblinking stare, lived on the shore of Lake Michigan in a small, quiet town called Petoskey, where he raised bees in his backyard and practiced ophthalmology.

Ehrlich's book had "a big influence on me," he says. He joined Ehrlich's ZPG movement in 1969, even before Ehrlich appeared on *The Tonight Show*, and bought copies of *The Population Bomb* by the case to hand out to friends and neighbors. He and his wife were ardent conservationists and committed community activists. Tanton had founded the Petoskey chapter of the Audubon Society and together with his wife co-founded a Planned Parenthood branch. He maintained a lifelong membership in the Nature Conservancy.

But the population control movement's emphasis on lowering the birth rate as the primary strategy for reducing population growth puzzled him. He'd seen the ecological value of other methods inside the beehives he'd been keeping since he was a teenager. Every fall, when the size of the bee colony reached its peak, the female worker bees would forcibly expel the male drone bees from the hive. They'd block the entrance to the hive, preventing the male bees from returning to its refuge. They'd drag any male bees still inside the hive to the precipice and throw them off the edge. Beekeepers such as Tanton would find the expelled drones' famished and frozen bodies with telltale chew marks on their wings. These were victims, Tanton presumed, of a brutal but necessary form of population control, a heartless mass eviction similar to the mass suicide of the lemmings.

The more he thought about it, the more he saw parallels between the beehive and human population. Wasn't the nation-state sort of like a hive, too, with its complex civilization enclosed within cozy borders? When the population of the bee colony exceeded the capacity of the hive to accommodate it, the bees took dramatic action. They evicted the dead weight and closed the borders. Didn't their behavior, he wondered, "raise questions about the human enterprise"?

Actually, it didn't. A honeybee is more like a cell in a body than an individual in society: it's a component of a larger whole. Many bees can't even feed themselves, and most play no role in reproduction. Evicting drones is nothing like deporting foreigners. It's like sloughing off dead cells. And a beehive is not like a nation. It's an enclosed and exclusive habitat. It's more like a private residence.

Hardin had used a similarly misleading metaphor to call for closed borders. In an influential 1974 essay, he likened the countries of the world to separate lifeboats adrift at sea. That was why, he said, wealthy nations had to close their borders to people from poor nations. Passengers from one lifeboat clambering aboard another could end up swamping it.

But nation-states are not isolated, self-contained units surrounded by impassable terrain. Except for a few remote island nations, they're connected to each other by land or by navigable waters. Human populations share the same more or less contiguous habitat. A more apt maritime metaphor would have likened the nations of the world to different parts of a single boat, and migration to passengers moving from one part to another.

Tanton prided himself on his emotional detachment, farsighted rationality, and moral integrity. Unlike Ehrlich, Tanton wasn't a charming man or much of an orator. He spoke in a flat monotone, his harsh views obscured by a mild, pedantic midwestern exterior. He was more of a moralist, the kind of person who considered it his ethical duty to scold someone who dropped a cigarette butt into a creek.

He was careful, in his public pronouncements, not to be openly hateful or bigoted. But privately, he had another reason to resist immigration. Tanton considered foreigners biologically alien.

He characterized them as separate species with distinct biological capacities: "*homo contraceptivus*" were those born in Europe and the United States, who generally had smaller families than did the newcomers, "*homo progenitiva.*" These outsiders, he claimed in private correspondence, would "bring their traditionally high fertility patterns with them." (In fact, immigrant fertility rates conform to local ones within a generation.)

Like Madison Grant and Henry Fairfield Osborn decades earlier, Tanton characterized intelligence as a biological feature passed down unchanged from generation to generation, in which education and opportunity played little or no role. That was why "less intelligent" people, as he wrote to a colleague, should "logically have less" children than the more intelligent ones. Even political cultures lay embedded within bodies, which immigrants would drag around with them like phantom limbs. "If through mass immigration, the culture of the homeland is transplanted from Latin America to California, then my guess is we'll see the same degree of success with governmental and social institutions that we have seen in Latin America," he wrote in a private letter.

Such deeply rooted differences between locals and foreigners were the reason why, for Tanton, migrants had to be barred from entry into the nation. "How could *homo contraceptivus* compete with *homo progenitiva*," he wondered in private correspondence, "if borders aren't controlled?"

Tanton thought the population control movement needed to do something about it. "I started writing to ZPG," he recalls, "and saying, 'If you're interested in numbers of people, what difference does it make whether they're born here or they move here?'"

They couldn't see any. ZPG invited Tanton to write background papers to help the group develop positions on immigration policy. Environmentalists in the population control movement shifted their focus from too many people *in the world* to too many people *in the nation*. The National Wildlife Federation resolved to support restrictions on immigration as part of its environmental platform. Top environmental thinkers such as the ecologist David Pimentel, the conservation biologist Thomas Lovejoy, the sustainable development advocate L. Hunter Lovins, the Blue Planet Prize–winning economist Herman Daly, and others served as board members or advisers to groups such as the Carrying Capacity Network, which called for an immediate moratorium on immigration into the United States.

Tanton quickly ascended to the upper echelons of the ZPG movement. He joined the board. He became friendly with Ehrlich and Hardin. By 1975 he was president of ZPG. The group called for slashing immigration into the United States by 90 percent.

For too long, Tanton wrote in a cover article for *The Ecologist* magazine, environmentalists had been overly focused on reducing the birth rate, allowing the "role of international migration in perpetuating population growth" to escape notice. The growing size of the human population "dwarfs the absorptive capacity of the few countries still willing to receive legal (and certainly illegal) immigrants," he wrote.

The only solution was to do as the bees did: evict the surplus and close the borders.

~

The July 4, 1977, edition of the *Washington Post* landed with a thud on millions of doorsteps, with an explosive two-thousand-word exposé of India's population control program on its front page.

The reporter who'd written it had traveled through small villages across India, reporting on the front lines of the global war on out-of-control population growth. He'd visited a small village of mostly poor Muslim families about two hours south of New Delhi called Uttawar, where electrical service had been abruptly cut off in 1976. The villagers had no idea why, they told the *Post* reporter, until a small contingent of police officers and local politicians arrived to explain.

They'd turned the electricity off on purpose, they said. And they'd turn it back on only if the local men agreed to undergo vasectomies.

The people of Uttawar were so reluctant to get vasectomies that they managed without electricity for two months. Finally the authorities could wait no longer. At three o'clock one November morning, a booming loudspeaker awakened the sleeping villagers.

A police officer announced that the village had been surrounded by armed guards equipped with gasoline. "Do not try to run," the officer said. "We will shoot you and burn the village down. All men and boys come out quietly."

Terrified men and boys trickled out of their homes and into a waiting line of trucks and buses, which whisked them away under the dawn sky. They hustled the boys to a local jail. The men were sent to an outdoor clinic where clinicians took a scalpel to their groins.

The *Post* report added to a steady trickle of tales of abuse and violence committed in the war against Malthusian catastrophe. The Indian government had deprived those without proof of sterilization of ration cards and land allotments. It paid people to snitch on neighbors who avoided the surgeries. Botched vasectomies conducted in ad hoc open-air clinics like the one in Uttawar killed over two hundred Indian men.

The violations did not stop at the Indian border. In China, seven-months-pregnant women were forced to undergo abortions. In the United States, a dozen states considered passing laws requiring women on welfare to undergo sterilization. African American teenagers were forcibly sterilized at federally funded family planning clinics.

Outraged feminists attacked Ehrlich as the mastermind behind the widespread human rights abuses. Groups such as Women Against Genocide showed up at Ehrlich's public appearances, handing out leaflets titled "Bomb Ehrlich."

Ehrlich was forced to backtrack. He blamed the excesses of his call to action on the uncertainties of the science on which it had been based. The book "had its flaws," he allowed, but only because "science never produces certainty." At the same time, he admitted that the book hadn't really been rooted in science at all, despite the fact that he had long capitalized on his gravitas as a scientist and a Stanford professor to promote it. *The Population Bomb*, he said, had been a "propaganda piece," aimed at galvanizing interest in environmental protection. "I was trying to get something done," he said.

~

By the time the population control movement crashed, the demographic trends that had fueled its ascent reversed themselves.

Thanks to people working together to develop and share new technology, improve education, and modernize societies, the pull of the demographic transition—by which people have fewer babies when the death rate falls—asserted itself over the postwar Malthusian blip. The U.S. birth rate had started to decline in 1955, even as millions of Americans panicked about a coming population explosion. By 1972 it had fallen below the level recommended by ZPG. Global population growth had peaked too. Demographic transition theory reestablished its authority. A 2009 *Nature* paper called it "one of the most solidly established and generally accepted empirical regularities in the social sciences."

Activists concerned about the state of the environment turned to less contentious ways to protect nature than railing about poor people's reproductive habits. Those concerned about poverty turned to the incrementalist socioeconomic development programs they'd discarded earlier. While population control efforts remained an important—and highly contested—part of international development programs, the population problem faded from the headlines. Within a handful of years, writes the historian Thomas Robertson, "despite isolated pockets, population had fallen off the national agenda almost entirely."

Quietly, out of the public eye, ecologists started to rethink the infamous raft of studies and observations that they'd trumpeted as proof of incipient Malthusian collapse. The depraved rats in their Baltimore enclosure; the lichen-killing reindeer hordes of St. Matthew; the clouds of smog above Los Angeles; the predictions of impending famine: all came to be seen in a new light.

They resurrected the work of the University of Chicago ecologist Warder Clyde Allee, who had documented the positive effects of population density, and even the negative effects of low population density, back in the 1930s. Allee spent his summers at the marine biological lab at Woods Hole, where he walked along the coast of Cape Cod for hours, collecting creatures he found along the shore. He had noticed how, in his glass-bottomed bucket, the snakes and starfish he collected tended to clump together tightly and how, in the eelgrass that washed up with the tide, the starfish never appeared alone but always in groups. He wondered whether there was some reason why.

In a series of what would later be hailed as "unambiguous" experiments in an "impressive range of taxa and ecosystems," Allee found that crowding actually improves the survival of individuals. A group of goldfish, he found, can survive in a liter of poisoned water for 507 minutes; a single goldfish, in contrast, can survive the same water for only 182 minutes. A clump of worms snuggled together can survive UV radiation 1.5 times longer than a single worm that slithers off on its own. Sea urchins and frogs that live close to one another lay higher densities of fertilized eggs than those spread out over long distances, and their eggs develop faster. Allee even figured out some of the mechanisms by which crowding improves life for individuals, for example by allowing aquatic creatures to benefit from the protective chemicals secreted by others, which are otherwise too diluted to have an effect.

Bringing individuals together, in other words, produces varying forms of social cooperation, which help individuals survive and thrive. It is why fish form schools in the sea, why birds flock, why mammals travel together in herds, and even why, Allee suspected, newcomers form neighborhoods when they settle in strange new cities. Ecologists called it the "Allee effect." Experts from a range of fields recognize the phenomenon, too. Modern neuroscientists call it the "hive mind."

In other words, ecologists who'd adopted the simple Malthusian calculation had left out an important part of the equation. They considered the costs of each additional human being: more mouths to feed, more cars on the road, more stress on natural resources. But they hadn't considered the benefits.

The Allee effect shed new light on the tragedies of the Kaibab and St. Matthew Island. The collapse of the ungulate populations in these predator-free locales shocked because they'd been characterized as places of easy living where the ungulates should have thrived. But the deep canyons encircling the Kaibab tablelands and the towering cliffs and Arctic seas surrounding St. Matthew also meant that the deer and the reindeer could not move.

The reindeer had trampled their lichens not because their crowds were too dense but because they were marooned. The Kaibab and St. Matthew Island were not paradises for the ungulates, despite being predator-free. They were prisons.

New understandings about the benefits of social cooperation explained why the catastrophic effects of population growth that Ehrlich and the other Malthusian ecologists warned about failed to materialize. Ehrlich's argument that the world would run out of nutritious food had relied heavily on Vogt's predictions that countries such as Mexico would soon be unable to sustain their growing populations. But he hadn't taken into account that working cooperatively together allows people to innovate more efficient agricultural and other technology. Instead of a growing population outstripping the food supply, the food supply expanded. Between 1944 and 1963, Mexico's production of wheat grew by a factor of six.

The innovative capacity of groups of people working together created new technology and collective actions that controlled environmental problems such as California smog, which the Malthusian ecologists had presumed would inexorably worsen as the state population continued to grow. Instead, catalytic converters and regulations on vehicle emissions lifted the blanket of smog over California's valleys, even as the number of people and cars in the state continued to climb. While technology and social cooperation were no cure-all, they formed an important counter-weight to the costs of population growth.

The scientific basis for Ehrlich's claims about the social depravity of crowds had started to disintegrate almost as soon as he'd made them. In a 1971 paper that described the scientific research underlying *The Population Bomb*, Ehrlich had written that crowding "may increase aggressiveness in human males." He'd cited as proof the research of Jonathan L. Freedman, the social psychologist he'd encouraged to conduct studies in humans of the social depravity Calhoun had found in rats, although at the time Freedman had yet to publish the results of his research. When he did, about a year later, he reported just the opposite, a reversal similar to Harry Shapiro's after visiting Pitcairn Island. "Crowding does not have a generally negative effect on humans," Freedman wrote, and "what effects it does have are mediated by other factors in the situation." In a book based on his research, he celebrated the bene-fits of high-density living.

Later research even mitigated the finding of social depravity among Calhoun's rats, jammed into their enclosures. Calhoun had later seen Allee effects, too. In one experiment, crowding had led the rats to innovate new ways of building burrows. Inspired, a *National Geographic* writer who'd visited Calhoun's lab wrote one of my all-time favorite children's books, *Mrs. Frisby and the Rats of NIMH*, in which a mouse family is saved by the superintelligent rats bred at the National Institute of Mental Health.

But while the social panic about out-of-control population growth diminished, deflated by demographic shifts and political scandal, the movement to make migration as difficult and deadly as possible not only persisted, it grew.

~

In 1979 the Michigan ophthalmologist John Tanton spun off ZPG's immigration committee into a new group aimed entirely at restricting immigration called the Federation for American Immigration Reform. He and his allies created a slew of associated organizations, all focused on cracking down on the flow of migrants into the country. Within a few years, Tanton's antimigrant network included the Center for Immigration Studies, an anti-immigration think tank; NumbersUSA, an anti-immigration lobby group; U.S. English, a group that opposed the bilingual education that newly arrived immigrants relied on; and Social Contract Press, a publishing outfit specializing in anti-immigrant literature.

Tanton's goal was to "make the restriction of immigration a legitimate position for thinking people." For a while, liberal advocates were broadly sympathetic to his economic and environmental arguments against migrants. In the 1980s and '90s, elements on both sides of the political spectrum aligned both for and against immigration, with corporate interests and their partisan allies broadly aligned in favor of immigration and labor unions and their partisan allies arguing that immigrants drove down wages and had a negative impact on the environment. Garrett Hardin and Anne

Ehrlich served on the board of Tanton's Federation for American Immigration Reform.

Like Ehrlich, who primed his readers and viewers to accept the necessity of authoritarian measures, Tanton gently helped his supporters disregard those who might call his antimigrant positions "racist." For too long, he'd tell them, environmentalists had been averse to discussing the truth about immigration because of the "seamy history" of "xenophobia and racism" that surrounded it. But those who truly cared about the planet and its people knew better. "We're not anti-immigrant," he'd say, "just as someone on a diet is not anti-food." It's just that "we have to address the finitude of resources," and because foreigners were unfailingly fecund and backward, "we can't do that by moving people around."

It worked, for a while. Then in 1988 the *Arizona Republic* exposed Tanton's private remarks describing foreigners as a lasciviously breeding subspecies. The civil rights group the Southern Poverty Law Center listed Tanton and his organizations on its damning list of hate groups. The conservative commentator and George W. Bush adviser Linda Chavez, who'd served as president of NumbersUSA, resigned in protest, decrying Tanton's "anti-Catholic and anti-Hispanic bias." "Any hope of significant liberal support," the *New York Times* noted, "vanished."

The gap between the two movements birthed by the population panic widened. Brower and a faction of anti-immigration activists within the Sierra Club brought the growing tension to a head by proposing that the group explicitly adopt an anti-immigration policy as part of its environmental platform. "Overpopulation is a very serious problem," Brower explained, "and overimmigration is a big part of it." A coalition of feminist and civil rights activists objected. After a series of bitter fights that ended with Brower's resignation from the Sierra Club board, the proposal was defeated.

The break with the mainstream environmental movement freed Tanton to reach deeper into other circles. Eco-nativists, worried about the impact of foreign peoples on the environment, flocked to the Tanton network. So did social nativists worried about the degrading effect of foreign cultures, eugenicists concerned about foreigners' impact on the gene pool, and white supremacists worried about the diminishment of their political power. Tanton invited their leaders into his home, supported their leading thinkers, and disseminated their ideas and writings through his publishing company.

That included what *Politico* would later call the "bible of the alt-right," a dystopic 1973 French novel called *The Camp of the Saints*. In the novel "swarthy hordes" of Indian migrants, described as "grotesque little beggars from the streets of Calcutta" who eat feces, invade France, force white women to work in brothels, and engage in orgies involving men, women, and children. The far-right French leader Marine Le Pen kept a dedicated copy in her desk. The former Breitbart chairman Steve Bannon considered the novel prescient and visionary. He suggested that the flow of migrants across the Mediterranean Sea would create a similarly horrific social meltdown. He called it an "almost *Camp of the Saints*–type invasion."

∼

Tanton's organizations reconstructed the Fortress America that Grant and Osborn had built. They successfully led efforts to defeat a 2007 bill that would have provided legal status to millions of people who'd crossed the border without permission; they mobilized opposition that successfully defeated the DREAM Act, which would have provided legal status to those brought to the

United States without permission as children; they helped write the notorious "show-me-your-papers" law implemented in Arizona, under which the failure to show valid immigration documents became a state crime.

After the war in Syria began in 2011, a new social panic about migrants erupted, creating another political opening for Tanton's network. The Trump administration tapped people from Tanton's organizations to oversee immigration policy. The office tasked with helping immigrants whose visas and citizenship applications had been denied or delayed would be overseen by Julie Kirchner, a former executive director of the Federation for American Immigration Reform. The administration's panel on election integrity would be led by the group's legal counsel, Kris Kobach. The head of the organization's polling firm, Kellyanne Conway, would become one of the president's top advisers. The federation's director of lobbying, Robert Law, would serve as a senior policy adviser to the Trump administration's U.S. Citizenship and Immigration Services, where he'd recommend that the government reduce the number of refugees it admitted and end the practice of automatically granting citizenship to people born in the United States.

In 2018 thirty-two of the thirty-four representatives in Congress who had earned an A-plus rating from NumbersUSA won reelection; the former Alabama senator who had been the subject of effusive press releases and awards from the organization, Jeff Sessions, ascended to the office of the attorney general. Sessions's aide, Stephen Miller, rose to become one of President Trump's chief policy advisers and speechwriters. He crafted the administration's immigration policies, including a 2017 executive order banning people from several majority-Muslim countries from traveling to the United States at all.

～

Tanton died in the summer of 2019. By then his antimigrant ideology had reached the highest echelons of global power. In the White House and the halls of Congress, a gaseous mixture of three hundred years of outdated scientific ideas wafted freely.

Antimigrant politicians and advocates described their vision of biological inheritance as Grant and Osborn had, as if complex traits passed down unchanged from generation to generation. "You have to have the right genes," Trump said. "I have great genes," he announced. "I'm proud to have German blood," he added. "Great stuff." He had a "genetic gift," he said, for business. His Treasury secretary agreed: "He's got perfect genes." The Trump family, one of Trump's sons said, subscribed to the "racehorse theory" of inheritance, which "places a high value on bloodlines."

They referred to the inferior biology of people of African descent, as Linnaeus had. "Some people," noted Trump adviser Steve Bannon, in reference to black people shot by police, are "naturally aggressive and violent." "Laziness is a trait in blacks," Trump said. "Some people cannot, genetically, handle pressure," he added. "Go out in nature," said a Republican nominee for Illinois representative, "and you don't find equality anywhere . . . I don't believe in this doctrine of racial equality."

They implied that mixing biologically distinct peoples disrupted the natural order, as early twentieth-century eugenicists had. "'Diversity' is not our strength," one of President Trump's national security officials wrote. "It's a source of weakness, tension, and disunion."

They argued for the restoration of a Linnaean vision of nature, in which biologically distinct peoples lived separately in geographically distinct locales. "Defined, ethnically and racially homogenous homelands" were the goal, as one white nationalist and Trump supporter put it. "We can't restore our civilization with somebody else's babies," said Steve King, a Republican representative from Iowa.

Antimigrant politicians in the United States mostly refrained from discussing environmental problems of any kind. But antimigrant politicians in Europe, echoing Aldo Leopold, Garrett Hardin, and the other neo-Malthusian ecologists, openly denounced migrants for the environmental burden they supposedly exacted. The antimigrant politician Marine Le Pen planned to remake Europe as "the world's first ecological civilization" by closing the borders to migrants. "Nomadic" people, she claimed, "do not care about the environment." "Borders are the environment's greatest ally," a spokesperson from her party added. Unhinged advocates agreed. Twenty-eight-year-old Brenton Tarrant from Australia, who aimed to repel migrants fleeing climate change, took matters into his own hands. In the spring of 2019 he slaughtered fifty-one worshippers at two mosques in Christchurch, New Zealand. A twenty-one-year-old from Dallas named Patrick Crusius allegedly claimed that "if we can get rid of enough people, then our way of life can become more sustainable." In the summer of 2019 he drove 650 miles southwest to the border to stop what he called a "Hispanic invasion." He opened fire at a Walmart in El Paso, Texas, killing twenty-two in the third-deadliest mass shooting the state had ever seen.

"Let them call you racists," Bannon said in a speech to an antimigrant party in France. "Let them call you xenophobes. Let them call you nativists. Wear it as a badge of honor . . . History is on our side."

～

Antimigrant advocates like Bannon imagined a version of the past that biologists had championed for centuries. It was one in which peoples lived separately in long isolation, adapting to their distinct landscapes and differentiating from each other. It was one in which migration's proper role was to rid ecosystems of excess individuals, and in which the movement of peoples across landscapes and biological borders presaged ecological doom. It was one in which modern migration, by bringing biologically distinct people together, disrupted the natural order.

That vision had been laid out over centuries. But when scientists finally turned their attention to probing its details, they found that most of it was wrong.

7

## HOMO MIGRATIO

Just how people had moved around the planet remained an
open question until well into the late twentieth century.
The resurrection of Darwin's ideas after the Second World War
elevated the possibility that at some point in the distant past there'd
been some ancient migration from a common source. But while
Darwin had argued that all our ancestors had originated in Africa,
the migration itself remained shrouded in mystery. Darwin never
explained how we'd distributed ourselves to disparate corners of
the planet. People presumably could have walked out of Africa into
the contiguous land masses of the Old World. But how they had
reached the heights of the Himalayas, the depths of the Amazon,
the frozen tundras of the Arctic, and the remote islands of the
Pacific remained unclear.

If anything, decades of scientific research since Darwin had
heightened scientists' awareness of the impassability of geographic
borders in the absence of modern navigational technology. Biologists described how long isolation had differentiated us from each
other, recapitulating the geographical borders that separated us
into biological distinctions between the bodies of people who

lived on different continents. They highlighted the dangers in crossing those borders. They documented the nefarious motives and disruptive impacts of movement, from suicidal zombie lemmings to depraved starving Indians. All of it underlined the rarity of migration in our past.

And yet the puzzling fact of ancient human habitation in places to which they couldn't possibly have walked persisted. Botanists, anthropologists, and geneticists put forward a series of theories to explain it.

~

The islands of Polynesia lie scattered across the vast Pacific, lava-spewing tips of volcanoes thousands of miles from the continents, surrounded by tens of thousands of feet of water on all sides.

It had taken centuries of effort for European explorers to make their way there. Only the most masterful European adventurers had had the skills to successfully navigate to these distant islands, thousands of miles from home. The English explorer James Cook, who'd navigated around the islands of the South Pacific in the late eighteenth century, had capitalized on the latest navigational techniques and technology. He used charts and magnetic compasses. He made complex measurements. With newly developed marine chronometers, he calculated the amount of time that lapsed between the sun rising high in the sky back home in England and over his own sails, which he then analyzed using spherical trigonometric methods to figure out how far west they'd sailed.

And yet, when he arrived at the Pacific's watery outposts, he found that a cacophony of living things had preceded him. From Tahiti to Hawaii, the Pacific islands were fully inhabited and alive with thousands of species of plants, birds, and animals.

The people from one island to the next—even those separated by thousands of miles—appeared to share a kinship. At one point, Cook ferried a high priest from Tahiti with him across the South Pacific. The man had spoken to distant islanders less like a stranger than like a long-lost cousin, their languages mutually intelligible.

For Cook, it was as remarkable as discovering that his dog could speak to his plants. Where had these people come from, and how had they successfully colonized all the Pacific islands, where expanses of sea eclipse land masses by a factor of five hundred to one? The logical conclusion, that prehistoric peoples had navigated from the continents over thousands of miles of open ocean, hopping from one remote island to the next and spreading their cultural and linguistic habits, seemed impossible. Long-distance migration was widely known to be an exceptional feat, requiring uncommon prowess and advanced modern technology. "How shall we account for this Nation spreading it self so far over this Vast ocean?" Cook scribbled in his journal.

In the late nineteenth century, Stephenson Percy Smith, the son of English civil servants who'd settled in New Zealand, tried to solve the mystery. Perhaps the Polynesians' migration to the Pacific could be explained by their superior racial heritage. According to his "Aryan Polynesian" theory, the prehistoric settlers of the Pacific had actually been Westerners, too. He pointed to linguistic evidence purporting to show that Polynesian languages originated in Sanskrit and other Old World languages. He referenced Polynesian peoples' "charming personalities," which showed that they must have derived from a "common source with ourselves from the Caucasian branch of humanity."

Peoples in Polynesia had an alternative theory. The Maori in New Zealand recounted that their ancestors had arrived in Polynesia from a land to the west that they called Hawaiki.

They'd carried with them the crops and animals they'd need to settle the islands, such as pigs, dogs, and fowl. And they'd done it centuries before Cook arrived, traveling not in a modern vessel using advanced technology but in a canoe called the *Taki-tumu*, sailing across thousands of miles of open ocean, against the prevailing winds and currents, with no charts or compasses to guide them.

Other canoe-traveling migrations had followed, they said, including a massive armada of canoes known as the Great Fleet. European travelers such as the Portuguese explorer Ferdinand Magellan, who landed in the Marianas Islands in 1521, had marveled at the speed and navigability of the locals' canoes. The French explorer Louis-Antoine de Bougainville had been so impressed by their vessels that he'd called Samoa the Navigator Islands. Cook, too, had noted that the place names, artifacts, and languages of the peoples of the Pacific bore eerie resemblances not just to one another but also to those in Asia.

The Norwegian adventurer Thor Heyerdahl, who arrived on the Polynesian island of Fatu Hiva in 1936 to study botany, did not accept the Great Fleet theory. Heyerdahl's theory of Polynesian migration drew on the legend of Kon-Tiki Viracocha, a chief who, according to legend, floated to Polynesia from Peru on a balsa wood raft.

A floating object could essentially drift from the coast of the Americas westward across the Pacific, with no particular means of navigation, on the prevailing winds and currents, Heyerdahl thought. Trade winds that circle the earth around the equator blow westward at a steady thirteen miles per hour. The cold waters of the Humboldt Current flow north along the western coast of South America, then head due west toward the equator at an average speed of about eleven miles per hour. Heyerdahl envisioned Aryan mariners traveling along the coast of the Americas who got swept up in storms or through navigational error got blown off course. The prevailing winds and currents would have deposited this race of "white gods" amid the islands of the Pacific.

Such a raft journey would explain how people with what Heyerdahl considered Stone Age technology had accomplished what, for Europeans, was a high-tech feat of migration. It would also explain the curious presence of sweet potatoes in Polynesia, Heyerdahl thought. European explorers had first encountered the sweet potato in the Americas. Perhaps people had floated from Peru to Polynesia, as Kon-Tiki had, and had brought American sweet potatoes with them.

Heyerdahl's theory maintained the fiction that borders, in which migration is a by-product of modernity, like electricity and telephone service. If the ancestors of the Polynesians had arrived by a Kon-Tiki-style raft, there hadn't been any purposeful migration at all. They'd arrived by accident.

∾

Heyerdahl's Kon-Tiki theory presupposed a fantastic journey. The raft would have had to encounter miniscule specks of land after drifting across five thousand miles of open ocean. It would have been like dipping your hand into the sea and inadvertently touching a dolphin swimming by. While the Great Fleet theory and its notion of Asians migrating to Polynesia failed to convince the Western scientific establishment, for many of Heyerdahl's colleagues, the Kon-Tiki-raft theory appeared equally improbable. In 1946 Heyerdahl approached a group of influential American anthropologists for support for his theory. They scoffed. One said mockingly, "Sure, see how far you get yourself sailing from Peru to the South Pacific on a balsa raft!"

To Heyerdahl, who'd grown up camping in Norwegian snow caves and climbing mountains with his Greenland husky, this sounded like a proposal. He did not know how to swim. He had never sailed, nor spent any significant time on the water. (If he had, he later said, he would have known that "you couldn't cross the ocean in the Kon-Tiki.") But he had faith in his theory: it made too much sense to not be true.

He mustered a small crew, procured some balsa wood in Ecuador, and set off for the port of Callao in Peru, where he built a nine-log raft outfitted with radio equipment and a painting of Kon-Tiki on its rudimentary sail. He persuaded the U.S. military to provide sleeping bags, field rations, suntan lotion, canned goods, and navigational and radio equipment.

On April 28, 1947, a tugboat launched Heyerdahl's little wooden raft into the brisk waters off the Peruvian coast. The crew aboard the raft, young Norwegian scientists, sent regular dispatches to the Norwegian embassy using a radio transmitter powered by a hand-cranked generator. Stories tracking their progress appeared in newspapers around the world. They "feel safe," the *New York Times* assured its readers on July 7, and were no longer fearful that "every twist, groan and gurgle inside the raft meant it was coming apart." They'd been "caught in a gale," the *Times* reported the next day. The day after that the storm died down, but the crew lost their "fine green parrot, Mauri," and were "desperately fighting sharks, tuna and dolphins." In just a few hours, the crew pulled seven encircling sharks and two tunas out of the water, and an octopus had washed aboard.

Finally, after 101 days of drifting in the Pacific, the Kon-Tiki raft ran aground on an uninhabited atoll in the Tuamotus, a group of French Polynesian islands. They'd drifted 6,900 kilometers, carried by currents and winds from the western shore of South America to these Pacific islands, just as Heyerdahl suspected the prehistoric settlers of Polynesia had.

Upon his return to Europe, Heyerdahl wrote a book about the Kon-Tiki journey. The account was wildly popular. Publishers translated it into fifty-three languages. A couple of years later, Heyerdahl produced a film about the journey, which won the Academy Award for best documentary in 1951. The idea that ancient migration had consisted of a series of mishaps delighted the viewing public. Scores of other explorers followed in Heyerdahl's footsteps, building their own rafts to re-create the accidental drifts that they believed had populated the Pacific.

The most salient objection to Heyerdahl's Kon-Tiki theory, that a raft set adrift on a random course in the Pacific would be unlikely to intersect with its widely dispersed specks of land, had been proved baseless.

In 1963 the historian Andrew Sharp published a devastating critique of the Great Fleet theory of Stone Age voyages to Polynesia from Asia. Polynesian canoes had had no keels, no metal fastenings, nor any of the other features of European vessels. That excluded them from the ranks of those vessels technologically capable of the journey, he wrote. The Great Fleet was just a legend used to justify insecure locals' territorial claims. Hawaiki was not a real place but a mythical site, like the Garden of Eden and Atlantis. Local stories about earlier migrations had included some obviously fraudulent content, after all, like people hitching rides on albatrosses and floating on pieces of pumice. The case was clear, he wrote: there'd been no purposeful prehistoric migration into the Pacific from Asia.

If anyone had succeeded in propelling themselves across the globe without modern know-how, they could have done it only by serendipitous misadventure. For the millions of people enthralled by the Kon-Tiki adventure, the case of the peopling of remote Polynesia was closed: they'd gotten there by mistake.

∾

The University of Pennsylvania anthropologist Carleton Coon, president of the American Association of Physical Anthropologists, went further. He claimed that there'd been no prehistoric migrations at all.

According to his theory, humans hadn't commonly originated in Africa. Weaving together the fragmentary evidence available in the fossil record, he argued in his 1962 book, *The Origin of Races*, that each of the human races had emerged and evolved separately.

Populations of now-extinct *Homo erectus* had dispersed across the planet and slowly evolved into modern *Homo sapiens* independently on each of the five continents, Coon argued. These disparate evolutionary journeys explained how the bodies of the five continental races had become as biologically distinct as generations of scientists had maintained. "Each major race," Coon wrote, "has followed a pathway of its own through the labyrinth of time." Over the course of millennia, they'd been variously "molded in a different fashion to meet the needs of different environments." If so, there was no reason to suggest that there'd been any migrations at all before the modern era.

Scientific belief in biologically distinct racial groups remained widespread. In 1950, in the wake of revelations of Nazi crimes in the name of biological distinctions between peoples, a top agency of the newly formed United Nations issued a statement officially condemning race as an ideological concept, with no basis in biology. But when the agency asked leading scientists to sign on to the statement, they'd balked. Even those sympathetic to the cause of antiracism were reluctant. "I need but mention the well-known musical attributes of the Negroids and the mathematical ability of some Indian races," the British primatologist W. C. Osman Hill protested. Dismissing the biology of race was "wishful thinking," another added. Clearly the mental capacities of the races differed, the evolutionary biologist Julian Huxley said, pointing to the "rhythm-loving Negro temperament" and the "shut-in temperament" of Native Americans. "I fear that I would not like my name to appear on the document," Huxley added. The evolutionary biologist Theodosius Dobzhansky, whose work in genetics had been instrumental in resurrecting Darwin's theory of natural selection, felt the statement went too far, too. In the end, 83 of the 106 prominent anthropologists and geneticists asked to sign the statement refused. Those who did sign it, Coon claimed, just gave it "lip service," then privately "tore it apart."

Coon's theory about the human past explained not just racial difference, which scientists accepted as a point of biological fact, but also the still-powerful fantasy of a racial order. That some racial groups had accrued far more political, economic, and social capital than others was plain to see. Instead of attributing that result to the political, economic, and social policies that lavished resources on some racial groups and deprived others, Coon's theory chalked it up to evolutionary history.

Each race's peculiar history of evolution from *Homo erectus* to *Homo sapiens* had occurred at different rates and at different times, Coon said, such that "each had reached its own level on the evolutionary scale." The Caucasoids—white Europeans—had evolved into *Homo sapiens* before any of the others, which was why they were "more evolved." Australian aborigines had only recently become humans, which was why political authorities were correct in treating them as primitive. Black Africans, whom he called Congoids, had "started on the same evolutionary level" as Europeans and Asians but then "stood still for half a million years." They'd become human so recently that they were essentially two hundred thousand years less evolved than white people, Coon wrote.

Influential scientists lauded the book. Coon's theory, while "highly speculative," the anthropologist Frederick Hulse noted, was "really comprehensive." In the pages of *Science* magazine, the Harvard evolutionary biologist Ernst Mayr lauded Coon's book as "bold and imaginative" and of "major scientific importance."

Because the idea of a racial order in nature did not contradict mainstream scientific thought at the time, those scientists who objected to Coon's theory, such as Dobzhansky, did so on technical grounds. The fossilized remains Coon used as evidence in his argument presupposed the biological distinctions for which he argued, Dobzhansky pointed out. Because they assumed a sedentary prehistoric world, archeologists categorized fossils discovered in far-flung locales as zoologically distinct. They dubbed the archaic human remains found in Indonesia *Pithecanthropus erectus*, for example, and those found in China *Sinanthropus pekinensis*, as if they couldn't possibly be the same species because they'd been found so far from each other. (In fact, both *Pithecanthropus erectus*, or "Java Man," and *Sinanthropus pekinensis*, or "Peking Man," would turn out to be specimens of a single roaming species, *Homo erectus*.) Coon used such fossils, with their suggestive nomenclature, as evidence for millennia of isolation between their descendants. It was like proving the distinction between a pretty thing and a beautiful thing because one had been called "pretty" and the other "beautiful."

Plus, Coon's theory conflicted with what scientists understood about evolution. If groups of *Homo erectus* had indeed been marooned on their own continents and evolved separately from each other, they'd have been unlikely to all evolve into the exact same species, as Coon's theory suggested. More likely they'd diverge into five different ones. Cases of so-called convergent evolution, in which separate lineages evolved in ways that rendered the same result, like marsupials in Australia and birds in Asia both evolving ducklike bills, were relatively rare. Evolution wasn't a train track leading engines inexorably to the same destination. Coon's theory required not just one Black Swan event but a handful of them, and all with the same result. The chances of that happening were "vanishingly small," Dobzhansky noted.

Even if all five marooned populations of *Homo erectus* had collectively evolved into the exact same species, if Coon's theory were correct, they would have had to practice strict isolation from one another. If they'd moved around and interacted, the inevitable battles and love affairs between them would have resulted in generations of, say, Mongoloid-Congoid babies and Caucasoid-Aboriginal babies, in whose bodies the biological differences of their ancestors would quickly fade. Despite having evolved into the exact same species, ancient peoples would have to have behaved as if they had not, keeping their distance from their fellow *Homo sapiens* elsewhere as if they'd been infected with some deadly contagion. Coon must have imagined that newly evolved *Homo sapiens* "practiced racial segregation during their wanderings," Dobzhansky noted dryly.

Civil rights activists condemned Coon's theory as a racist fantasy. The Anti-Defamation League published a pamphlet condemning his claims. His colleagues in physical anthropology called a special meeting to censure his work, forcing Coon's resignation as society president. Segregationists, meanwhile, rejoiced in Coon's notion of barely human Africans and primitive aboriginals. They disseminated his theory in newspapers and in their own pamphlets. Coon himself maintained an active correspondence with and provided scientific feedback to leading segregationists, such as Carleton Putnam, whose book on the biological backwardness of African people inspired a young Ku Klux Klan enthusiast, David Duke.

Coon brushed off his critics, characterizing them as "Pavlov's puppies." Dobzhansky, he sniffed, was a "stuffed jackass" waging a "campaign of defamation." Experts who told the "truth about race" such as Coon were being "persecuted," his supporters said.

~

Decades would pass before scientists recovered the long-suppressed history of human migrations and shattered the myths of a sedentary past and a racial order. Until then, each new discovery suggesting migratory movements would be crammed into the old paradigm. It finally began to crack the year after Coon's book came out, when a couple of experimental biologists at Stockholm University, peering at the cells of chick embryos through an electron microscope, spied some strange fibers.

Those fibers, tucked inside the cells' mitochondria—worm-shaped structures that generate cells' energy—turned out to be DNA, the same stuff that coiled inside the cell's nucleus. But unlike the DNA in the cell's nucleus, which mixed and reassorted with one's partner's DNA in unpredictable ways during reproduction, the DNA inside a mitochondrion was kind of a solitary outpost. It contained just a few dozen genes and quietly traveled through the generations solely through the maternal line, from mothers to babies, unaffected by the confused scramble of reassortment. Its order changed solely through a steady drip of random mutations.

That meant, UC Berkeley's Allan Wilson realized, that differences between DNA sequences could describe the passage of time, like the depth of a sedimentary layer or the number of rings on a tree. Genetic changes that accrued at a predictable rate acted as a kind of stopwatch, recording the number of times they'd passed through the generations, the way a corroded phrase at the end of a game of Telephone might. He started using this insight in the 1970s to compare the sequences of genes and proteins between different species, seeing if he could pinpoint the moment of their divergence.

Until then such questions had been the province of paleontologists and anthropologists, who pieced together the story of the deep past based on shards and scraps of artifacts, fossils, and other clues. Wilson's "molecular clock" technique fractured their findings. Paleontologists had concluded that chimps, gorillas, and humans had been evolving separately for about 15 million years. Wilson's research suggested that they'd parted ways only 3 to 5 million years ago.

In the late 1980s, Wilson and his colleagues at Berkeley, Rebecca Cann and Mark Stoneking, persuaded a few hundred pregnant women with recent ancestry rooted in different continents to donate their placentas for a study of their mitochondrial DNA, to find out how long their ancestors had been evolving separately. The researchers froze the placentas, ground them up in a blender, then spun them in a centrifuge a few times, extracting a clear liquid containing pure mitochondrial DNA.

Most experts agreed that regardless of whether our ancestors arrived on their separate continents via an ancient migration or emerged there, as Coon argued, the peoples of Africa, Asia, the Americas, and elsewhere had been evolving separately, behind their impregnable geographic borders, for at least a million years.

That's not what this mitochondrial DNA showed. According to the geneticists' analysis, the 147 women of different racial and continental backgrounds had shared common ancestors as recently as two hundred thousand years ago. If true, the long period of isolation that scientists had presumed for centuries didn't exist. The peoples of the world had emerged from a common ancestor so recently that they had had far less time to differentiate than previously believed. And they had migrated during prehistoric times in a far speedier and more extensive fashion than anyone had imagined. In just a few hundred thousand years, people had catapulted themselves into every last corner of the planet.

The scientists poetically dubbed their shared ancestor whose unbroken line of daughters had bequeathed them her mitochondrial DNA "Mitochondrial Eve." Mitochondrial Eve explained scientists' decades-long failure to locate clear biological distinctions between us: they didn't exist. As the evolutionary biologist Richard Lewontin established in the early 1970s, the variation between racial groups accounted for less than 15 percent of the total genetic variation across the entire species. Much more variation existed between individuals—whether of the same race or not—than between the races.

After centuries of allegiance to sedentist myths, commentators viewed the notion of a mass migration out of Africa with suspicion. Critics complained that Wilson and his colleagues "simply didn't have the training" to untangle the complexities of the human past. "African migrants" could never have successfully colonized the entire planet, the paleoanthropologists Alan G. Thorne and Milford Wolpoff wrote in a 1992 *Scientific American* article.

~

Research by scientists such as the population geneticist Luigi Luca Cavalli-Sforza supported the migratory history suggested by Mitochondrial Eve. Cavalli-Sforza called that ancient exodus the "Recent Out of Africa" migration. He, among others, incorporated the new DNA evidence with the way skulls had changed, how pathogens, languages, and cultures had evolved, and a raft of other archaeological evidence to prove that we had indeed migrated out of Africa just hundreds of thousands of years ago.

Cavalli-Sforza's work forced the fact of our recently shared African origins into mainstream acceptance. But his theory left other central planks of the sedentist paradigm intact.

The premodern migrations he described occurred under exceptional and short-lived circumstances. The way Cavalli-Sforza imagined it, the journey out of Africa had been a dispersal into empty land, motivated by the allure of unoccupied territory. Our earliest ancestors evolved in Africa in a world of vast, unpopulated spaces, "new, pristine environment[s]," and "virgin territory." They spilled out of Africa the way a pool of water expands to fill an empty container. Colonizers set out from Africa to settle new places and founded new colonies, which hatched more colonizers to settle more new places, founding more colonies, and so on until all new places were studded with human habitations.

At that point, the historically unique conditions that compelled our prehistoric ancestors into motion vanished, the migratory process came to its natural conclusion, and the potent barriers to migration imposed by geography and culture rose up once more. Cavalli-Sforza's and his colleagues' assumption about what happened next was the same as other scientists' since Linnaeus: thousands of years of stillness, until modern technology artificially lowered nature's barriers to our movement.

This assumption had been baked into his own research. Cavalli-Sforza reconstructed the historical relationships between peoples by analyzing their DNA. But to reconstruct the path of prehistoric journeys out of Africa, he didn't collect DNA from a random cross-section of people across the globe. Instead, he focused on a particular subset—indigenous peoples, in particular those who spoke their own languages and lived within well-defined geographic borders—whom he imagined had remained in the same places where their ancestors had deposited them since time immemorial. He pieced together the story of their ancestors' movements out of Africa by measuring the relatedness of the long-immobile descendants they'd left behind.

Local communities targeted by his team of scientists had not been pleased. The presumption that the subjects whose blood the scientists sought were no more than "isolates of historical interest" to be pierced, classified, and filed away in gene banks rankled. In Central African Republic, an angry farmer accosted Cavalli-Sforza as he drew blood from a local child. "If you take the blood of the children, I'll take yours," he warned, brandishing an ax. The World

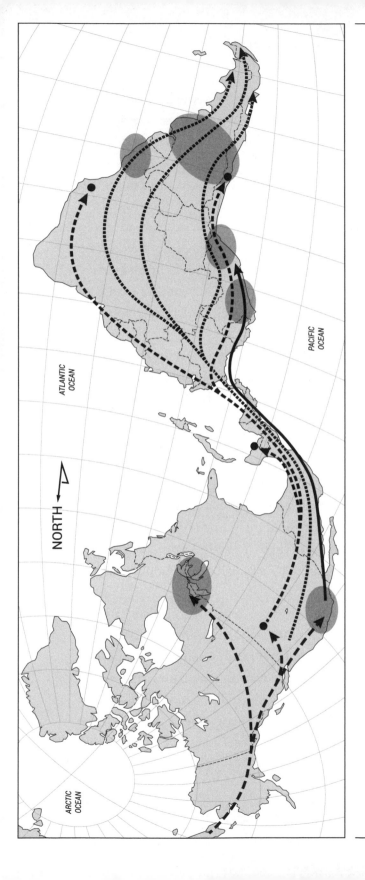

NORTH

ARCTIC
OCEAN

ATLANTIC
OCEAN

PACIFIC
OCEAN

## Peopling the Americas

Until the discovery of ancient DNA in petrous bones in 2014, population geneticists depicted prehistoric human migration as a single dispersal into unpopulated lands, followed by a long period of immobility until modern transportation overcame geographic barriers to movement. New data from ancient DNA suggests a history of nearly continuous migrations. This map depicts the multiple waves of prehistoric migration into the Americas.

- ⬤ Basins of human population

- ● Archaeological sites

### Genetic lineage

- ⟶ 4,200 years ago
- ⇢ 9,000 years ago
- ⇠ 14,000 years ago
- ⇠ 16,000 years ago

*Source: Cosimo Posth et al, "Reconstructing the Deep Population History of Central and South America," Cell 175, no. 5 (November 2018): 1185–97.e22.*

Council of Indigenous Peoples dubbed Cavalli-Sforza's work the "Vampire Project." The Third World Network, an NGO, called it "totally unethical and a moral outrage."

Some of Cavalli-Sforza's colleagues objected to his strategy, too, arguing that the groups of people who met his criteria may not have been the isolated and immobile people he presumed. They, too, might be a mix of migrants from different places with a peculiar and checkered history of trade, exchange, conquest, and cultural collision. Perhaps, in other words, the migration out of Africa had been followed not by millennia of stillness but by still more migrations. Cavalli-Sforza's method of swooping in to extract blood samples as if his subjects' ancestors had no migratory history to speak of would miss it entirely.

"I am very troubled," one scientist told a reporter from *Science* magazine. "By sampling that way you bias the results." Instead of attempting to untangle genetic relationships between groups of people living in different places (or even asking them about their own migratory histories), Cavalli-Sforza's method simply presupposed them in advance. If those groups of people happened to be as mixed and migratory as their ancestors, his strategy was almost as misleading, the anthropologist Jonathan Marks later wrote, as "asking whether lawyers are more closely related to architects or to accountants."

But apart from a few grumblings, Cavalli-Sforza's method stood. Scientific ranks closed around the Recent Out of Africa theory in the first decade of the twenty-first century, a century and a half after Darwin had first proposed a common origin in Africa. Documentary films, museum exhibits, and magazine articles popularized the new story of the human past that DNA technology had helped reveal. Many used the metaphor of a tree. The trunk represented ancient peoples of Africa, from whom we'd all evolved. Each population that walked out of Africa into another continent appeared as a branch, reaching out into the distance.

In fact, there was no direct evidence that migration had essentially stopped after the dispersal out of Africa, as the metaphor suggested. The strands of DNA in the ancients' cells that might have recorded their movements had rotted and decayed, along with their long-buried bodies, millennia ago. But most presumed they'd stayed still. Branches, after all, do not grow back together.

∿

Hints that the past was neither isolated nor sedentary appeared in 2000 with results from the Human Genome Project, a multibillion-dollar program to sequence the human genome.

The sequencers had found barely any difference in any of our genes. According to their results, a paltry 0.1 percent of the sequence of the 3 billion nucleotides strung together on our of DNA differed from any one person to the next. Men and women, the short and the tall, the red-haired and the black-haired, the tongue-curlers and the droopy earlobed and the color-blind, all shared an almost identical sequence of nucleotides in their DNA. Our species had not diverged into separate branches at all. Human beings, President Bill Clinton proclaimed in the White House ceremony announcing the results, were 99.9 percent the same, "regardless of race."

Relatively speaking, we hardly have any genes at all, the results showed. Ever since the days of Weismannism, scientists had believed in the commanding power of biological inheritance. Molecular geneticists had described DNA as a master molecule directing the development and functioning of our bodies as if by dictatorial fiat. The geneticist Richard Dawkins had likened human bodies to "lumbering robots," manufactured by the sequence of nucleotides in our DNA. Genes played such a central role in our health and behavior, scientists thought, that decoding their sequence would cure cancer and revolutionize the economy. Our gene sequence would tell us "what we 'really' are," recalled Jonathan Marks.

Scientists had expected the human genome to include at least one hundred thousand different genes. They knew that the genome of a millimeter-long nematode worm has around twenty thousand genes. If genes controlled our bodies and behaviors the way many suspected they did, highly complex *Homo sapiens* would surely have many more, they figured. But as the project proceeded, scientists had had to recalibrate their estimates of the number of genes in the human genome. In 2001 they predicted it might carry not one hundred thousand but perhaps thirty thousand genes. In the end, researchers who analyzed the number of genes revealed by the Human Genome Project sequence found just around twenty thousand—about the same number as the lowly worm. Whatever distinctions we noticed among us could not be encoded in our biology in any simplistic fashion, passed down intact from generation to generation. We didn't have enough genes to spell the difference.

"No one could have imagined," said one, "that such a small number of genes could make something so complex." Fewer than ten genes, his colleague added, could separate humans from mice.

Studies of the genetics of our fellow primates made the biological borders between us appear even more ethereal. Ernst Mayr had distinguished between species in which biological changes from population to population were abrupt, with each group having character combinations distinct from that of others, and species in which such changes were continuous, shading imperceptibly from one to the next. Chimpanzees and honeybees were of the former type. Our genes revealed us to be of the latter.

Chimpanzees, primatologists had found, live in closed groups that don't mix with other groups, even when their habitats overlap. That's reflected in their genetics. Chimpanzees, gorillas, and bonobos all have more genetic diversity within their ranks than we do. The genetic distance between two populations of chimps, geneticists found, is four times greater than the genetic distance between people living on different continents. The isolation of their populations from one another allowed them to differentiate. But the same hadn't happened to us, despite the fact that we are far more numerous and widely distributed than they are. A history of migration and mixing explains why.

Still, confronted with the new genetic evidence, many scientists felt compelled to hang on to the myth of Linnaean borders between us. Some felt, like an earlier generation of race scientists, that racial boundaries might yet be found and scientists simply had to look harder for them. In a 2002 study, for example, population geneticists decided to sidestep the subjective bias of people self-reporting their own racial categories. Presuming that biological fault lines between the races existed and that the computer could find them "objectively," they fed genetic data from 1,052 different people into a computer program called STRUCTURE and asked it to find the genetic borders between them. Since, as genetic evidence suggested, migration had made the pattern of variation between people continuous and graded, this was sort of like asking a computer program to analyze the number of colors in a sunset. The result hinged entirely on what number the program was told to find. If the researchers asked for three groups, STRUCTURE would sort the data into nonsensical, non-race-based groups such as "people from Europe," "people from Africa," and "people from East Asia, Oceania, and the Americas." When they asked for six groups, the software sorted the data into people from each of the continents plus a separate group consisting of people who lived in the mountain valleys of northwestern Pakistan known as the Kalash. It could even sort the data into twenty different groups. Still, when it came up with the five continents after being told to divide the data into five groups, investigators proclaimed victory. In an interview with the *New York Times*, the study's lead author, Marcus Feldman, said the study had confirmed the popular conception of race.

Other scientists agreed. "Looked at the right way," the Imperial College evolutionary developmental biologist Armand Marie Leroi commented in a *New York Times* op-ed, "genetic data show that races clearly do exist." STRUCTURE had had "no knowledge of the population labels" in common use, added the Harvard geneticist David Reich, but had clustered people into the five groups that "corresponded uncannily well to commonly held intuitions about the deep ancestral divisions among humans."

Reich's notion of "race" allowed for more nuance than it did in conventional usage. For him, race referred to a genetically related population group, not the broad conglomerations that Linnaeus had defined by skin color and continental origin. Still, in a 2018 *Times* op-ed, he mostly elided the distinction. "Differences in genetic ancestry that happen to correlate to many of today's racial constructs," he wrote, "are real."

Myths about Linnaean-style biological differences between human racial groups continued to seduce medical professionals. In a 2016 study, for example, half of white medical students claimed that black people's skin was thicker than white people's skin. That false belief, which correlates with medical professionals' inability to accurately assess black people's pain, is likely implicated in the pregnancy-related deaths of black women, which occur at a rate three to four times higher than in white women. Other scientists found racial categories to be scientifically convenient, regardless of whether they were biologically relevant. In medical genetics studies, for example, scientists continued to group geographically and genetically diverse populations such as Koreans, Mongolians, and Sri Lankans together as "Asians" and Moroccans, Norwegians, and Greeks as "whites," just as Linnaeus had recommended centuries ago.

Similarly, maps depicting human genetic variation portrayed continental populations as separate entities with visible discontinuities between them. One such map, which Cavalli-Sforza and colleagues published in a 2009 *PLOS One* paper, represented the populations of Africa with red dots, those in the Americas with pink dots, those in Europe with green dots, and those in Asia with orange dots. A similar map appeared on the cover of one of Cavalli-Sforza's books.

Those colorful distinctions conformed to racial categories more than to the actual relationships between the data sets. The range of genes present in people living in Europe or Asia was not distinct from the range of genes present in people living in Africa, as such maps suggested. Because the peoples who had originally migrated out of Africa to settle the other continents composed a subset of the entire population of Africa, their descendants' genes were a subset of those present in the peoples of Africa. A more accurate pictorial representation of the range of genes present in different populations might color the African continent using an entire palette of pigments but use just a randomly selected but overlapping fraction of pigments for Europe, Asia, and elsewhere.

People who had faith in the myth of biological race and a racial order found sufficient scientific evidence to back up their beliefs. One popularly cited statistic from the Human Genome Project noted that people are 99.9 percent the same "regardless of race." That didn't mean that a consistent 0.1 percent genetic difference defined racial groups. The differences between individuals did not, in fact, fall along racial boundaries. But the locution left open that possibility. Given that we share 98.7 percent of our DNA with chimpanzees, and 90 percent with mice, a 0.1 percent difference between races is not necessarily insignificant. "After all," one observer noted to the race scholar Dorothy Roberts at a conference, "dogs and wolves are nearly identical at the genetic level, but the difference between a dog and a wolf is huge."

Cavalli-Sforza's color-coded maps had anti-immigrant and white supremacist commentators crowing with delight, including on the popular white-supremacist website VDARE. "Basically, all his number-crunching has produced a map that looks about like what you'd get if you gave an unreconstructed Strom Thurmond a paper napkin and a box of crayons and had him draw a racial map of the world," one VDARE writer noted, referring to a notorious pro-segregation U.S. senator from South Carolina. Cavalli-Sforza and his colleagues, he concluded, had "largely confirmed the prejudices . . . of nineteenth century imperialists." These color-coded maps, another VDARE advocate wrote in the *San Diego Tribune*, painted a pretty clear picture. Each race was as clearly distinct from the others as pieces of fruit in a bowl. "What does that sound like to you?" the writer asked, suggestively.

Policies that failed to recognize racial biology were "pseudo-scientific," another white supremacist website, the Daily Stormer, opined. "Science is on our side," added the founder of a white nationalist group. Some even posed as genetics experts themselves. In early 2019, for example, the Maryland congressman Andy Harris, who held a degree in medicine from Johns Hopkins University, took a meeting with an advocate to discuss the "number of sequenced genomes for research," as a congressional aide put it. That person had no training in genetics. He was a fund-raiser for white-supremacist outfits.

Just as they had decades earlier when the UN asked them to officially condemn the concept of race in human biology, experts in the field retreated from the political implications of genetic research. "Many geneticists at the top of their field say they do not have the ability to communicate to a general audience on such a complicated and fraught topic" as the biology of human difference, a *New York Times* article noted. When an organizer with the American Society of Human Genetics attempted to schedule a panel on the political misuse of genetics research, she "found little traction," the *Times* noted. David Reich, for example, refused her invitation to lead a public discussion on the issue.

Plus, the hard boundaries of race conformed to a larger view of history that had embedded itself in the public mind. As anyone who'd seen the pictures of the tree representing our population history knew, each continental race had traveled on a separate bough to its destiny, independent of the others. That's what the DNA revolution had revealed—at least until geneticists got their hands on some petrous bones.

≈

The petrous bone is named after the Latin word *petrosus*, for "stonelike and hard." It's the part of the skull that encases the tissue-lined labyrinth of the inner ear, allowing us to interpret vibrations as sound. It's the hardest and densest bone in the mammalian body.

It has also protected bits of DNA from the forces of degradation for tens of thousands of years, a fact that geneticists who examine ancient remains happened upon around 2014, when they analyzed a few bony fragments that included petrous bone. Until then, they'd generally stuck to pulverizing femurs and tibias in their search for old DNA, on the theory that weight-bearing bones were the most likely to have retained it intact. As a result, they'd found little ancient DNA in the skeletal remains they examined, besides those that had been preserved under ice or in deep caves.

The discovery of the petrous bone revolutionized paleogenetics. Inside its bony swirls is what one paleogeneticist called the "mother lode" of ancient DNA. In 2010 the genomes of five ancient people were published; by 2016, there'd been three hundred; by 2017, more than three thousand had appeared. The work of incorporating the new data rushing out of paleogeneticists' labs into our understanding of migratory history has only just begun. But already paleogeneticists such as Sweden's Svante Pääbo and Harvard's David Reich, among others, have revealed a backstory of ancient migrations that is far more complex than what Cavalli-Sforza and others extrapolated from modern-day DNA.

The Out of Africa journey had been cast as a dispersal into vast empty spaces. But when our ancestors walked out of Africa, new data from ancient DNA revealed, they moved into lands where other peoples already lived. These now-extinct archaic humans had beaten us there, having migrated out of Africa themselves some 1.8 million years ago. When our ancestors encountered them, they did what migrants do everywhere: they had babies with the locals, a process of mixing that allowed bits of their DNA to enter ours. About 2 percent of the DNA in modern-day peoples in Europe and Asia traces back to the migratory collision with Neanderthals; and around that proportion of DNA in people now living in New Guinea and Australia traces back to the Denisovans, a group of ancient humans discovered through genetic analyses. A Denisovan gene that allows people to survive at high altitude now resides in the DNA of people living in Tibet.

Ancient peoples, after their arrival in Eurasia and the Americas, hadn't stayed put either, ancient DNA reveals. Some migrated back to Africa, endowing their modern-day descendants in eastern and southern Africa with genes from Eurasian peoples. Others migrated to India, joining streams of ancient migrants from Central Asia, the Near East, and the Andaman Islands, all of them leaving their genetic fingerprints behind. Ancient migrants who arrived in

Southeast Asia later set off for Madagascar. Those who'd migrated into the Americas picked up and left for Europe.

Geographic barriers—open oceans, mountain ranges—had not barred their wanderings. Nor had a lack of modern navigation technology. Ancient migrants washed over even the remotest regions of earth, and they'd done it successfully more than once. For years, scientists had figured that ancient peoples had migrated into the forbidding Tibetan plateau 15,000 years ago. According to new DNA analyses, they'd also migrated there 62,000 years ago.

No freak accident deposited unsuspecting people on the remote islands of the Pacific. Ancient peoples had been so determined to settle the Pacific Islands for so long that despite the navigational and technological challenges that the journey entailed, they'd successfully made it there in three distinct waves before Captain Cook arrived, as archaeological, linguistic, and genetic evidence shows.

Patterns revealing genetic relationships among far-flung populations suggest other unexpected journeys. The five-thousand-year-old remains of a farmer buried in southern Sweden turn out to be genetically related to people living in Cyprus and Sardinia today. Modern-day Native Americans turn out to share genes with the Chukchi people of northeastern Siberia, suggesting their ancestors' migrations from Asia into the Americas and then back again. The ancients roved to and fro to such an extent that even the most seemingly homogenous of their descendants—modern western Europeans, say—could not claim any long period of isolation and differentiation, as much as some might have liked to. The homogenous ancestral population that commentators such as Madison Grant and others imagined never existed. Several genetically distinct groups of people migrated into the region and variously mixed and melded with one another. From what paleogeneticists can piece together, they included dark-skinned hunter-gatherers, farmers with dark eyes and fair complexions, and another group of farmers with light hair. The western Europeans of today are hybrid descendants like the rest of us.

The past, in other words, is "no less complicated than the present," Reich notes. We weren't migrants once in the distant past and then again in the most recent modern era, with a long defining period of stillness in between. We've been migrants all along.

The image of the tree, with its separate branches representing the continental populations, suggested that continental populations had diverged, each evolving separately from the others as they reached off into the distance. But geneticists have found no evidence of such divergence. The seeming homogeneity within today's continental populations and races—the similar skin tone of northern Europeans, the straight hair of East Asians—is not the consequence of some long unbroken line of unchanging ancestry but the momentary result of an ongoing process of migration, differentiation, and melding together again.

Sometimes when two separate branches of a tree rub together in the wind, slowly removing a layer of bark, the layers of tissue growing underneath start to fuse together. As the conjoined branches thicken, bark growing around their wounds, they become a normal branch like any other; the immune fighters, microbes, and nutrients that once pulsed through the circulatory system of each branch separately now flow through the fused branches as one physiological entity. Botanists call the process "inosculation," from the Latin for "little mouth." It can happen between the branches of one tree, or between branches of separate trees.

The result, a braided tree with branches sprouting from its trunk and then merging back together again, is like a river with streams flowing in and out, winding apart and then reuniting.

If our past is a tree, it is this special kind of tree. Our ancestors migrated, met, merged, then migrated again. We continue to do the same today.

Linnaeus named our species *Homo sapiens*, Latin for "wise man." A more apt name might have been *Homo migratio*.

~

Pius "Mau" Piailug grew up half submerged in water. He'd been born on a single-square-kilometer speck of coconut trees known as the Micronesian island of Satawal, played in tide pools as an infant, and learned to sail at the age of four. Friends said his rippling back muscles recalled the shell of a hawksbill turtle.

He looked out over the low-lying bow of the *Hōkūle'a*, a sixty-two-foot double-hulled sailing canoe, as it sliced through the deep blue waters of the Pacific. The *Hōkūle'a* had been crafted to conform to eighteenth-century illustrations of traditional Polynesian vessels, drawn by Captain James Cook's crew. In it, Piailug would re-create the ancient migrations that had peopled Polynesia.

Linguistic, archaeological, and ancient DNA evidence has shown that prehistoric people migrated from Southeast Asia into the Pacific in at least three distinct waves. First, people crossed from China to Taiwan and the Philippines. Then they covered the open ocean to reach Vanuatu and Samoa. Finally, they reached the remotest islands of Polynesia, such as Hawaii and Easter Island. They hadn't come from Peru, and they hadn't arrived by accident.

The anthropologist Ben Finney estimates that over the millennia of prehistoric migrations into Polynesia, upward of a half-million migrants likely lost their lives at sea. But *Homo migratio* pressed onward regardless. Experts now widely recognize their migration as "arguably the most expansive and ambitious maritime dispersal of humans across any of the world's seas or oceans," as a 2016 paper in the *Proceedings of the National Academy of Sciences* put it.

*Hōkūle'a*'s passage across 2,700 miles of open ocean between Hawaii and Tahiti would require navigating two different trade wind belts, the windless doldrums, and equatorial currents and countercurrents that steadily push vessels off course. Piailug and his crew would have to dodge hurricanes, typhoons, and blustery squalls with winds that could reach up to thirty knots, and pass by active volcanoes spewing smoke and flames, surrounded by submerged boat-killing reefs.

Most modern mariners that attempt the voyage set off equipped with the latest navigational aids: powerful engines for when the wind dies, GPS devices and chart plotters to keep track of their course in the featureless ocean, satellite phones and other telecommunications to call for help. Even with all that, there are no guarantees. During one attempted crossing in 2017, two sailors encountered a squall that killed their engine and damaged their mast. They were lost at sea for five months. When they were finally rescued, they had drifted thousands of miles off course.

Piailug would use neither charts nor modern instruments of any kind. He'd rely solely on the traditional navigation techniques that ancient migrants might have used.

"Wayfinding" involved using stars, ocean swells, and behavioral observations to keep track of speed, distance, and position. It allowed mariners to locate their vessels on the open ocean even as winds and currents and waves battered them to and fro. It required making thousands of observations every day, of the position of the sun, moon, and stars, and the subtle changes in the behavior of birds and fish, which shifted depending on their distance from land. Sometimes Piailug would lie down on the canoe's floor to absorb the feel of the ocean swells, from which he could detect invisible bodies of land in the distance.

Wayfinding could take a lifetime to learn. Piailug had been taught by his grandfather and father. Heyerdahl and the other

Europeans who'd intruded into the Pacific hadn't known about wayfinding, in part because practitioners were forbidden from sharing the quasi-religious practice with outsiders.

Between 1976 and 2009, the *Hōkūle'a* completed nine voyages using traditional wayfinding. It completed its journey from Hawaii to Tahiti in thirty-four days.

~

Kon-Tiki wasn't a total bust, though. Heyerdahl was right about the sweet potato. It had come from the Americas. But people hadn't brought the plant with them on an accidental drift from Peru to Polynesia.

The potato had made it across the Pacific on its own. In 2018 a survey of sweet potato DNA, including DNA from sweet potato leaves that had been collected in Polynesia by Captain Cook's crew and stored in the National History Museum in London, showed that the Polynesian sweet potato had started to evolve separately from American sweet potatoes about 111,000 years ago, tens of thousands of years before humans reached Polynesia. Most likely it made the journey afloat on the water or was carried by birds.

Human migration is not exceptional. Long isolation did not differentiate our species into separate races. Feats of navigation are not the sole province of "white gods" from the West. The oceans can be crossed by canoe.

And humans aren't the only ones who move across the landscape, leaping over continents and oceans. Plants and animals do, too.

# 8

## THE WILD ALIEN

It's an hour before sunrise on an October morning when a few dozen baseball-capped birders, binoculars swinging atop their fleece sweaters, arrive on a grassy meadow on the coast of a narrow canal on the peninsula of Cape May, New Jersey.

Up to a million birds—peregrine falcons, sharp-shinned hawks, plovers called killdeer, snowy white mute swans, sea-diving northern gannets, and parasitic jaegers hatched on the Arctic tundra among them—can be spotted along the cape's narrow peninsula as they head south on their annual migrations. Sometimes cold fronts force them to congregate, forming rivers of birds that stream across the sky for hours.

The birders have woken up at an ungodly hour to enjoy the spectacle. They are connoisseurs of wild movements. But even they reflexively defend a natural order in which movement is reserved for a select few.

The morning sky bleeds from deep blue to a thin line of orange at the horizon. The birders scan the sky with their binoculars. Suddenly someone calls out. He's spotted something. "Flicker!" Everyone quickly turns to the patch of sky he's pointing to, readjusting their binoculars to find the migratory woodpecker he's identified. To my untrained eye, the flicker passing high overhead looks not unlike a black-crayon checkmark depicting "bird" in a child's landscape drawing, but the others murmur with awe and delight.

After a few moments, someone else spots a long line of the sea ducks called scoters flying low over the water. "This is what it's all about!" he yells triumphantly, punching his fist into the air. Later, when the group retires to a banquet hall for a buffet meal, red-cheeked and wind-tousled, someone mentions an observatory overseas, where birds can be seen migrating by at waist height, eliciting a collective gasp.

But as charmed as these bird-watchers are by the spectacle of moving birds, the movements of creatures they deem out of place do not enchant. Reeds known as phragmites grow in tall dense stands lining the edges of the canal and the bluff beside it. According to the fossil record, phragmites have been present in the United States for at least forty thousand years. A morphologically identical but more vigorously growing strain from Europe arrived around the early nineteenth century. The reeds grow deep and strong, displacing other wetland species like wild rice and cattails, but they perform useful ecological functions in the habitat, too, filtering and cleansing dirty water and providing material that can be used for thatched roofing, baskets, fishing poles, spears, and in Egypt, a little clarinet-like instrument called the *sipsi*. The stems can even be dried and ground into flour.

After the morning session of coastal observations concludes, the group passes by a stand of phragmites. Even the most expert among them cannot point to any specific harm the phragmites cause. But they condemn the reeds on principle, based on their foreign origins and conspicuous health.

"They are invasive," one woman explains to me. "It's a shame." The others grumble their agreement. "Look at how many seed heads they have," one says in disgust. "They're so hard to get rid of." If they'd been less polite, they would have spat on them.

The phragmites are just now filtering water and providing succor to the local wildlife. The sound of warblers called kinglets rummaging inside is audible. The birds, a woman next to me says, would be "better served by something native."

∾

Linnaeus, whose taxonomy first conflated wild species with geographic locales, had not delved into the question of where species originated or whether and how they'd moved into their present-day habitats. For him, species belonged *ipso facto* wherever he found them. And he inscribed that vision, depositing each species in its place, in the way he named them in his taxonomy.

Darwin's theory of evolution posed an early challenge to Linnaeus's vision. His notion that all species originated from a common source required that at some point in the past, species moved across the planet, even surmounting geographical barriers to arrive at their current habitats. Monkeys, which could not swim across oceans, had spread across the Old World as well as the New. Lizards had made it to outposts across the globe. Immobile wild creatures— beetles, trees, mollusks, and the like—had flung themselves from their common origins over unscalable mountains, unlivable deserts, and insurmountable seas.

Darwin imagined a series of Kon-Tiki-like accidents dispersing species over long distances. Seeds submerged in a bit of mud could get stuck between a bird's toes, for example, or encrusted along its feathers, before it took off for a long migration. The tiny shell of a mollusk could attach itself to the leg of a beetle, or adhere to the inside of a shell, before being swept out to sea by a storm. Rodents scavenging near coastal kelp beds could be carried away on floating rafts by ocean swells, allowing them to reach distant shores. Over time, he wrote, sufficient numbers of such accidental long-distance journeys could have dispersed species across mountains and oceans and deserts, depositing them on even the remotest shores.

Darwin had no direct evidence of these epic voyages, but he conducted experiments to prove that species could survive such journeys. He submerged seeds from eighty-seven different plant species in bottles of salt water, fishing them out after a few months to see if they still sprouted. He procured duck legs and dangled them in an aquarium to test whether freshwater snail hatchlings might cling to them. He forced seeds into the stomachs of fish, fed the fish to birds such as eagles, storks, and pelicans, then carefully extracted the seeds from the birds' droppings, which he germinated.

His findings suggested that 14 percent of all plant species produced seeds resilient enough to travel nearly a thousand miles.

He considered the peculiar assemblages of species on islands suggestive, too. Terrestrial species could in theory distribute themselves across continental land masses by walking, but could reach remote islands only through long-distance migrations over oceans. Indeed, he noted, islands were home to those species most likely to survive long-distance journeys. New Zealand, for example, had plenty of the plants and insects that could easily weather a Kon-Tiki raft ride, and none of the mammals and reptiles that couldn't.

Observers spotted the kind of happenstance conveyance Darwin envisioned in 1892, when a nine-thousand-square-foot floating island replete with thirty-foot-tall living trees was seen floating off the northeastern U.S. coast. They spied it again a few months later, about twelve hundred miles northeast. If it didn't fall apart before it reached a coast, such a floating island could facilitate the kind of long-distance colonization that Darwin suggested, ferrying seeds, insects, and other creatures along to some distant shore.

Scientists rejected Darwin's theory of long-distance dispersals regardless. Wild species moving around the planet in unpredictable and haphazard ways, irrespective of natural borders, violated the myth of a sedentary planet. While the fact of the wide distribution of wild species was difficult to square with their shared origins, that didn't justify abandoning the sedentist paradigm and speculating about random unpredictable events that could be neither tested nor predicted.

Many were willing to entertain even more fantastic theories, so long as they were consistent with a closed-border world. One popular theory posited that wild species had traveled from their common origins to their present distributions by walking across now-disappeared land bridges that once connected continents to islands and to one another. There was no "reasonable geological evidence" that such land bridges had ever existed in the places where enthusiasts imagined them, the evolutionary biologist Alan de Queiroz notes. Still, nineteenth-century writers drew fanciful land bridges on maps "willy-nilly wherever closely related species were found on both sides of a sea or ocean." One such map imagined a submerged land bridge wending over three thousand miles from southeastern Africa to the tip of India. Another connected West Africa to the eastern coast of South America; over it, a herd of elephants might stampede their way from Sierra Leone across the Atlantic directly to Brazil in a matter of days.

The conflict between Linnaean sedentism and Darwin's theory remained essentially unresolved for the better part of the twentieth century. The biogeographical theory that finally settled the clash emerged in the 1970s. It would stifle the history and promise of migration for decades thereafter.

~

The idea that the continents had once all been connected into a single whole was first proposed by the early twentieth-century German meteorologist Alfred Wegener, who had noticed how the continents' shapes could fit together like puzzle pieces. Through some mysterious process, he said, they must have split apart, the fragments drifting to their present locations.

Decades passed before anyone believed him, mostly because no known force on earth was powerful enough to pry apart large masses of solid rock and move the continents around over thousands of miles. He failed to find any convincing evidence before perishing, wrapped in a reindeer skin and buried under the Greenland snow, in 1930. But in the 1960s scientists discovered a geological force powerful enough to explain continental drift. The theory of plate tectonics is now taught in every introductory geology course.

Plate tectonics also resolved the dilemma of how species had spread across the planet in a sedentary world.

For hundreds of millions of years, the continents had been fused together as one, allowing the world's species to share a single contiguous land mass. That explained wild creatures' shared origins and biological commonalities. Then as the supercontinent fell apart—a process that continues to this day, pushing Plymouth Rock about fifteen meters farther west today than it was in 1620—the world's species must have been carried off in different directions. That explained their scattered distribution. Biogeographers call it the theory of "vicariance."

Vicariance obviated the need to imagine a past full of chaotic, unpredictable movements by flora and fauna across geographic borders. Any physical shift that had occurred in the past had transpired millions of years ago, without anyone moving a muscle or ruffling a pelt.

Wild creatures didn't cross oceans, mountains, deserts, or other geographic borders on their own. Deep underfoot, below the ponds, valleys, and glens in which mollusks, frogs, and snails lived, tectonic plates had imperceptibly shifted at a rate of about 100 millimeters a year for billions of years, unknown to the denizens above. Nobody actually moved anywhere much at all—the tectonic plates had moved for them.

Biogeographers started finding clues in geological history that explained the mysteries in species distributions they'd long pondered. Flightless birds with clearly shared ancestry lived in the far-flung continents of Australia, South America, and Africa. How had they dispersed so widely? Their common ancestor had likely populated each continent when the three had been connected. Hoofed ruminants lived in North America, where they'd evolved into moose and caribou, as well as in Asia, where they'd evolved into elk and reindeer. Their common ancestors probably populated the two continents when they were connected, too. Marsupials were nowhere to be found in India and Africa. Why not? They likely drifted off from the supercontinent before the ancestors of marsupials could climb aboard.

Biogeographers couldn't work out all the details of how geological forces had distributed species around. But they felt confident they would. Any number of geological changes could be called on to explain how species had been moved: the formation of a mountain range, slowly splitting a species into two; falling sea levels creating land bridges that allowed once-marooned species to colonize new territory.

Vicariance restored a "biological version of inertia," as de Queiroz put it. Biogeographers allowed that the accidental, long-distance dispersals Darwin imagined might have occasionally occurred, but otherwise they dispensed with migration as a coherent explanation of where species belonged and how they'd got there. The biogeographer Gary Nelson called Darwin's theory of long-distance dispersals the "science of the improbable, the rare, the mysterious and the miraculous." The very notion was "negative, sterile and superficial," the zoologist Lars Brundin added. It "offends the critical mind."

The few biogeographers who believed in long-distance voyages as a viable theory of history might as well claim that "some lucky humans" would "learn to fly," the paleontologist Paul Mazza wrote. In the story of the movement of species around the planet, biogeographers relegated random long-distance leaps to little more than "footnote acknowledgment." Such misadventures were "almost by definition random" and "hence uninteresting," a paper published in a 2006 issue of the *Journal of Biogeography* noted.

Biogeographers questioned not only the role of long-distance journeys in history but also whether wild species could even survive such voyages in the first place. Most animals could not, critics claimed. In one 2014 paper, a paleoecologist from the University of Florence described a little jackrabbit who'd been found floating on a bed of kelp set adrift by a storm, about forty miles off the coast of California. After a few days at sea, the bunny was half-dead from dehydration and heat exposure. The creature hadn't even made it across the dozen or so miles between California and the Channel Islands. No jackrabbit ever had.

～

In the history of nature envisioned by vicariance biogeographers, the movement of species had been so slow, passive, and imperceptible that active, long-distance wild movements could play no role in nature or in history. It underlined what many had known to be true since Elton warned of alien invaders after the Second World

War: plants, animals, and other creatures that crossed borders and entered novel territory were trespassers, invaders, and aliens who threatened the natural order.

The U.S. government has explicitly managed the national parks as oases from the ravages of alien border crossers since the 1960s, when it heeded the advice of conservationists such as Aldo Leopold's son, the zoologist A. Starker Leopold. He had recommended that the nation's national parks "preserve, or where necessary . . . re-create the ecologic scene as viewed by the first European visitors," which presumably was when Leopold suspected the long era of stillness ended.

The government extended those protections to the entire nation in 1999, when then-president Bill Clinton established the National Invasive Species Council, a body tasked with repelling "alien species" whose "seeds, eggs, spores or other biological material" were "not native to that ecosystem." After the terror attacks of September 11, 2001, guarding the nation against invasive species became one of the charter functions of the newly formed Department of Homeland Security, enshrining the business of policing natural borders into the national security infrastructure.

For years, conservation-minded people around the country had cleansed their gardens of alien species and joined native plant societies to champion the cause of the endangered natives, reflexively deriding newcomers like the phragmites lining the canals of New Jersey. Scientists joined the effort in the 1980s. Three new subdisciplines emerged—conservation biology, restoration biology, and invasion biology—all aimed at tracking the damages that border-crossing wildlife caused.

The pace of the onslaught was "unprecedented," one ecologist said. Already, over the last five hundred years, newly arrived species had come to dominate some 3 percent of the earth's ice-free surface. In many countries, these species composed 20 percent or more of the resident flora. California, England, Louisiana, and Chicago had been invaded by "German" wasps, "African" snails, "Chinese" crabs, and "European" mussels, one prominent invasion biologist warned.

In books with titles such as *Immigrant Killers*, *Alien Invasion*, and *Feral Future*, writers laid out the case against wildlife on the move.

According to the "enemy-release hypothesis," for example, intruders eluded native predators in ways that native species could not, giving them a dangerously unfair advantage; conversely, they preyed on native species in ways that native predators could not. In Hawaii, a local invasive species council noted, native species had lived "in relative isolation over . . . 70 million years," evolving in the island's "benign environment." These native inhabitants would be ravaged by an onslaught "nonnative, competitive" species from elsewhere, with their thorns, sharp hooves, toxic secretions, and carnivorous appetites.

Invasion biologists pointed to the intruders' growth as an indicator of their success in displacing native species. Argentine ants grew larger in areas they invaded than in their own native habitats, two Stanford biologists noted in a paper outlining the evolutionary impact of invasive species. Within just twenty years of arriving on the North American West Coast, fruit flies from Europe evolved altered wing sizes and extended their range from southern California to British Columbia.

The newcomers mixed with local species, which raised the specter that they'd contaminate local species with alien tissue. Mallard ducks hybridized with New Zealand gray ducks, Hawaiian ducks with Florida mottled ducks, sitka deer from Japan with reed deer from Great Britain. The interbreeding was "massive," Stanford biologists wrote in a 2001 paper in the *Proceedings of the National Academy of Sciences*.

They took ecological jobs from the natives, as in Britain, where American gray squirrels displaced the native red squirrels, and facilitated the arrival of more of their own kind, in a sort of chain migration. Introduced species formed "synergistic relationships" with other introduced creatures, according to a new preface that appeared in a 2000 reissue of Elton's 1958 book. As a result, if one appeared, there'd likely be many soon enough. The zebra mussel, for example, had enabled the arrival of a Eurasian water milfoil, a feathery flowering plant that lived underwater. In the arid hills of southern California, the hooves of introduced cattle destroyed the delicate crusts formed by lichens and mosses in the dry soil. That damaged the habitat of the native plants that the native checkerspot butterflies fed on. Alien plants thrived instead, pushing checkerspots to the edge of extinction.

Intruders such as the zebra mussel, which had arrived from Russia into North America during the nineteenth century and spread into the Great Lakes, marred boat hulls, blocked water pipes, and attached themselves to native clams, preventing them from getting enough food. Invasion biologists suspected that they'd cause the collapse of native clam populations. Invasives like purple loosestrife from Europe, growing in tall vigorous stands with prominent purple flowers, would displace native cattails and harm local wildlife. Municipalities spent millions of dollars trying to suppress it.

One invasion biologist calculated that wild species moving freely across the planet would ravage large swaths of ecosystems. The number of land animals would drop by 65 percent, land birds by 47 percent, butterflies by 35 percent, and ocean life by 58 percent. Based on assessments such as this, experts described newly introduced species as the second-largest threat to biodiversity in the United States. Invasion biologists tallied the net cost of biological invasions at $1.4 trillion, or 5 percent of the value of the global economy. The newcomers were the "mindless horsemen of the environmental apocalypse," the Harvard ecologist E. O. Wilson warned.

Given the stake, ecologists considered facilitating animal movement, whether on purpose or by mistake, to be so perverse and dangerous that they rejected the idea of moving species even to save them. Camille Parmesan, worried about the fate of checkerspot butterfly colonies, stood up at a scientific conference and suggested moving some threatened checkerspot colonies elsewhere. Her fellow ecologists erupted. The very idea overwhelmed them with horror and emotion, she remembers. "They accused her of playing God; of tampering with nature," a *Guardian* article on the ensuing hubbub reported. "Her approach would set off a whole new chain of problems."

If invasion biologists' and other scientists' concerns about the threat posed by species on the move sounded similar to those articulated about human migrants, they were. The corrective action required was similar, too. There'd be no relaxing of the borders, no welcome, no easing of assimilation for the newcomers. The intruders had to be eradicated. It was a "nasty necessity," Stanley Temple, an ecologist and science adviser to the Aldo Leopold Foundation, wrote in 1990.

∽

The team of scientists, wearing shorts and carrying axes and shovels, threaded their way through a tangle of jungle in the shadow of Mauna Loa, the world's largest volcano, which sprawls across the largest island of the Hawaiian archipelago, Hawaii. Inside this dense, humid forest, beneath the flaming blossoms of ʻōhiʻa trees, ancient biogeographical borders had been transgressed. The ragtag group of botanists, led by curly-haired Rebecca Ostertag and tall, wiry Susan Cordell, intended to do something about it.

The island's original inhabitants, some twelve hundred species of plants and animals that had made their way to Hawaii on their own steam, were a special bunch, with unique qualities that allowed them to survive the searing lava that periodically poured over the island. The changeling *ʻōhiʻa* could tolerate almost any kind of soil, including fresh lava flows. For millennia, it and other native Hawaiian species had lived in isolation from the chaos of the continents.

But then people started to come over, bringing cosmopolitan outsiders like pigs and dogs and rats and new diseases. They brought tree frogs from Puerto Rico, mongoose to control the rats, and ornamental plants to grow in their gardens. Birds ate their fruits and spread the seeds around in their droppings, which quickly bloomed under Hawaii's tropical sun. The island was soon overrun by sharp-elbowed newcomers, sucking up the island's nutrients and stealing its sun.

About half the flora clinging to the side of Mauna Loa, Ostertag and Cordell knew, were nonnative foreigners. For now, the old *ʻōhiʻa* trees still dominated the overstory, but that wouldn't be true for long. In 2010 farmers had noticed a strange fungus decimating the island's iconic *ʻōhiʻa* trees. Nobody knew what it was, exactly, but most presumed it was an alien newcomer, too. Soon it would take the *ʻōhiʻa* trees out. When it did, they'd be fully replaced by the outsiders. The young trees and saplings below bristled with foreigners. The forest ecosystem that the native Hawaiian species had forged thousands of years ago would vanish.

The botanists marked out four one-hundred-square-meter plots in the jungle. During a few brutal months of back-breaking labor, they destroyed every immigrant species they could find within its borders. They cut down trees with saws, then doused the stumps with deadly herbicide. They yanked shrubs and ferns, prying their roots out of the rocky ground. They set up giant funnels to capture any seeds that might rain down from above. Meticulously, they cleansed their plots of any detectable hint of nonnative tissue.

∼

The battle against invasive species presumed their movements into new habitats via global trade and travel to be historically and ecologically aberrant. But scientists didn't really know how far butterflies could fly, or whether wolves could surmount mountain ranges, or if crocodiles swam in ocean currents. For centuries, tracking animal movement had been a haphazard affair, relegated to the "margins of ecological research," as a 2015 paper in *Science* put it. Experimental methods, whether by design or by necessity, could rarely capture the scale of animals' wanderings across the landscape.

Like the British military's inadvertent discovery of bird migrations through radar, many dramatic and long-distance movements had been discovered by accident. European observers first understood that storks wintered in Africa, for example, after they'd happened upon the stork with a spear of clearly African origin pierced into its side. Nineteenth-century scientists considered the owls called saw-whets, which migrate thousands of miles every year, to be a "general and constant inhabitant of the Middle and Northern states," as one put it, until ornithologists happened to find thousands of birds in the sea after a freak snowstorm forced them out of the sky.

Even the now-famous migration of monarch butterflies from eastern North America to Mexico had been discovered serendipitously. In the 1930s Fred and Norah Urquhart, zoologists at the University of Toronto, had noticed the monarchs' disappearance every winter and their reappearance in the spring with tattered wings, as if they'd been on some long journey. They started gluing little tags on the butterflies' wings, reading "Please send to Zoology University Toronto Canada." Over the decades, only a few handfuls had been returned. Many arrived from points south of Toronto, but whether that indicated a butterfly flight or simply one swept up in the breeze remained unclear. The mystery would have continued except that in 1975 the pair of zoologists traveled to Mexico. During a hike into the mountains of Michoacán, they saw millions of monarchs coating the trees, their fluttering wings sounding like a waterfall. A pine branch laden with butterflies happened to crash down in front of them, a tiny paper tag affixed to one of the butterflies' wings revealing its northerly origins.

The popular "mark-and-recapture" method had scientists tying threads to birds' feet, as the nineteenth-century ornithologist John James Audubon did, or using Magic Markers to draw dots on butterflies' wings, as the twentieth-century butterfly biologist Paul Ehrlich did. Others used plastic ID tags, dyes, paints, and the like, or set up cameras that shot photos of animals that happened to pass by. By marking individual animals and then recapturing them later, their movements could be at least crudely inferred. But however animal trackers marked their wiggling subjects, mark-and-recapture methods allowed scientists to confirm only that subjects had moved where they thought they probably might. If a dotted butterfly or a bird with a thread on its foot evaded recapture, scientists were free to use their imaginations to determine what might have happened.

In one study, for example, Ehrlich had marked 185 butterflies and then set them free, returning to search for them a few days later. He found 97 of the marked butterflies flitting around just where he'd marked them in the first place. The other 88 eluded recapture. They could have moved beyond where he'd looked for them, but he assumed they had all perished. He concluded that checkerspots had a "remarkable lack of wanderlust."

Marking animals with tags that emitted signals that could be detected as they moved, like a bell on a cat's collar, circumvented the quandary of confirmation bias but introduced other problems. The tags could be heavy, which could disrupt the animal's behavior. They were expensive. "It cost 3,500 dollars," to tag a single animal, remembers one animal-tracking scientist. "You put it on the strongest animal" and hope for the best. They had limited battery life, so after a while they'd just stop sending signals, leaving scientists in the dark about where their slippery wearers had got to.

Some tried to conserve the tags' energy by programming them to ping only once a day or so, which gave only the most rudimentary outline of an animal's movements. But however scientists chose to triangulate between the weight, expense, and energy needs of their tags, to get any data they still had to essentially follow the animals around, capturing the signals on their receivers. Early efforts involved chasing behind the tagged birds by car, or helming light aircraft to fly slowly behind them so the little beeps and pings could be logged. "We'd have to physically go near the elephant," one animal tracker remembers, "fly over it, and locate it with the antennae on either side of our airplane until we caught sight of it. Then, by sight, we'd estimate where we were on a map and mark a little cross there. That's just the way it was."

The U.S. military had a much better system. MIT scientists had noticed that radio signals transmitted by the Russian satellite *Sputnik* increased and decreased as the satellite's orbit approached and then retreated. So the military started sending signal-emitting satellites into space. By the 1990s, its Global Positioning System of satellites emitted signals continuously, and there were so many that at least four could be detected at any place on earth at any time of day. In theory, animals outfitted with GPS tags could be tracked wherever they went on the planet, with no need to follow them around with a receiver. But fearful of aiding adversaries with navigational prowess, the Defense Department introduced an

# The long-distance journey of the sweet potato

Twentieth-century biogeographers posited that geological forces, not active movements, dispersed plants and animals into their present distributions around the planet. Molecular biology techniques suggest that plants and animals have undertaken a host of long-distance journeys in the deep past. This map depicts the long-distance journey of the sweet potato into Polynesia from the Americas.

SOUTH AMERICA

RAPA NUI (Easter Island)

EASTERN POLYNESIA

MARQUESAS

TUAMOTU

MANGAREVA

PITCAIRN

AUSTRAL

SOCIETY

HAWAII

COOK

SAMOA

FIJI

NIUE

VANUATU

KANAKY (New Caledonia)

AOTEAROA (New Zealand)

NEW GUINEA

Sources: Caroline Roullier et al., "Historical Collections Reveal Patterns of Diffusion of Sweet Potato in Oceania Obscured by Modern Plant Movements and Recombination," Proceedings of the National Academy of Sciences 110, no. 6 (February 2013); Douglas E. Yen, The Sweet Potato in Oceania: An Essay in Ethnobotany (Honolulu: Bishop Museum Press, 1974); Karl Rensch, "Polynesian Plant Names: Linguistic Analysis and Ethnobotany, Expectations and Limitations," in Islands, Plants, and Polynesians: An Introduction to Polynesian Ethnobotany: Proceedings of a Symposium, ed. Paul Alan Cox and Sandra Anne Banack (Portland: Dioscorides Press, 1991).

Area believed to be where the sweet potato was first introduced (according to Yen, 1974).

Primary prehistorical dispersal

Secondary prehistorical dispersal

Independent prehistorical dispersal (according to Rensch, 1991)

Late eighteenth- and early nineteenth-century dispersals by travelers unpredictable and erratic jitter into the signals, purposely degrading their accuracy. The GPS signals could be accurately interpreted only by receivers owned by the military. Everyone else got a uselessly faulty result.

And so for scientists as for the rest of us, the movements of animals remained mostly hidden. Even the ones who lived among us crept and crawled and darted around out of sight. Sometimes we'd notice, with a jolt, the faint trails they left behind—a few paw prints in the snow, an abandoned nest in the shrubbery— suggesting their passage. But usually, crossing paths with a wild animal, even common ones that lived around human habitations like deer or fox, was an occasion of surprise and delight.

A few weeks ago I saw a red fox in my driveway. It shouldn't have been surprising, since I had heard that a pair of foxes had moved into the neighborhood. But even though we'd been sharing our patch of suburbia for a few months, I had little awareness of their whereabouts. I froze in shock at the sight.

∿

The long, curving trunk of the highland tamarind tree arcs in a balletic stretch across the misty forest, thousands of feet above the sandy shores of Réunion, a volcanic island of less than one thousand square miles in the Indian Ocean. To find the tree, whose wood could be used to build fishing canoes and to roof houses, locals climbed three thousand feet up the pitched side of the volcano, until they glimpsed the trees' strangely twisting branches looming out of the fog as if part of some enchanted forest.

The otherworldly highland tamarind tree shares a striking similarity to another tree, which similarly lives on a volcanic island ringed by coral reefs: the koa tree, which can be found growing in the deep ash deposited along the slopes of Hawaii's volcanoes, smoky blue butterflies drinking nectar from its flowers. People in Hawaii use wood from the koa tree for their ukuleles and surfboards.

The likeness between the two species puzzled botanists for centuries. It seemed impossible that the Hawaiian koa and Réunion Island's highland tamarind shared any ancestry, as scientists could conceive of no movement that could have allowed one to seed the other. The two islands, separated by eighteen thousand kilometers of ocean, had no geographic or geological connection. They were as far away as any two specks of land on earth could be. No current or wind flow or migratory bird route connected them. Even if a seed had been somehow ferried across the ocean it would have been unlikely to survive the journey. The seeds are thin-walled and can't even float. Nor do they grow on seashores.

Botanists settled on two equally unsatisfying explanations for the similarities between the koa and the highland tamarind. Maybe the two trees shared no relation at all, in which case they'd somehow evolved to look exactly as if they did. Or maybe human migrants had picked them up and moved them around, although just who had done that, when and why, no one could really say.

Historical biogeography was full of such unsettled matters. Biogeographers stitched together the story of passive impercep-tible movements by associating geological events with species distributions, based on fossil evidence. When it wasn't possible, they came up with likely stories that made sense within their framework of a sedentary world.

Then, using the same molecular clock methods that had upended ideas about the timing and scale of human migrations, molecular biologists started testing those stories.

～

Scientists reported their findings on the genetic relationship between the koa and the highland tamarind in a 2014 *Nature* paper.

The highland tamarind, they'd found, had directly descended from the koa. Some Réunion Island tamarinds were, in fact, more related to Hawaiian koas than to each other. And the seed that connected the two had accomplished the epic journey between Hawaii and Réunion Island 1.4 million years ago, before *Homo sapiens* even evolved.

The genetic evidence meant that somehow the koa tree had traveled across eighteen thousand kilometers of ocean and colonized Réunion island. The koa's voyage was the longest single dispersal event ever recorded. And it wasn't the only one that molecular biology findings suggested.

Vicariance theory attributed the separation of monkey species into New World and Old World species to the opening of the Atlantic Ocean, which had slowly and passively separated the two lineages. But according to the findings of molecular biologists, the species hadn't diverged until 30 million years after that ocean emerged. The monkeys couldn't have been passively separated. Their ancestors must have crossed the ocean.

Rodents in South America, which according to vicariance theory had traveled over the Panamanian isthmus, arrived years before geological forces formed the land bridge connecting the two American continents. The rodents must have surmounted the sea.

Vicariance theorists presumed that the geological breakup of Gondwanaland, which split southern South America from Australia, and Madagascar from India, had gradually rent once-contiguous plant species. But that didn't match up with when the plant species diverged from one another. According to an influential 2004 study that one botanist dubbed "the last great gasp of the vicariance paradigm," the plants hadn't been carried on tectonic plates, either. They'd actively moved.

The molecular findings suggested a host of long-distance journeys in the deep past. Monkeys had made their way from the Old World to the New, at a time when that journey required crossing the Atlantic Ocean. Polynesian sweet potatoes, which had diverged from American sweet potatoes tens of thousands of years before humans could have carried them to Polynesia, colonized the Pacific on their own. Rodents catapulted themselves from North to South America before any land route existed. This kind of non-geological movement, irrespective of geographic obstacles, was exactly the kind of improbable, rare, and mysterious movement that Darwin had been talking about.

The koa tree's journey may have been a "giant fluke," de Queiroz commented, "but that's part of the message of a lot of recent biogeographic studies," he said. "Giant flukes happen."

~

As new molecular techniques recovered the dramatic story of animals' past migrations, other new technologies transformed scientists' ideas about how they moved in the present. The revolution in animal-tracking technology that made it possible began a few minutes after midnight on May 1, 2000. That was when the Defense Department stopped adding a jitter to its GPS satellites' signals, allowing them to flow uninterrupted to anyone in the world with a receiver. (They had figured out a way to selectively block the signals as necessary to deter adversaries.)

An $8 billion GPS technology industry sprang up, unleashing a blizzard of new products, including solar-powered GPS tags so small and light, they could be attached to the furry ear of a baby bear or the slippery shell of a sea turtle. New solar-powered GPS tags allowed people to track the once-undetectable movements of animals continuously, in real time, over their entire ranges and lifetimes. Animal-tracking scientists such as the ornithologist Martin Wikelski, who had grown up on a farm in Bavaria amazed by the local barn swallows that flew all the way to South Africa, quickly outfitted their roving subjects—cranes, dragonflies, oilbirds, and more—with the new tags. The new GPS data were augmented with observations from people around the world who were newly connected by social media. Whale-watchers shared observations with others in Iceland; bird-watchers uploaded millions of bird sightings via apps on their phones. By 2016, more than three hundred thousand birders had logged 11.8 million bird sightings around the world on one such app, eBird.

The results were stunning. "Every time we look," Wikelski says, "we find totally amazing new information . . . that turns around our knowledge."

Arctic terns logged 70,900-kilometer migrations, nearly twice as long as previous estimates. A few years later, another tagging experiment found that terns traveled nearly a third farther than even that. Jaguars in the Peruvian Amazon, whose ranges scientists had estimated to be about the size of Manhattan based on studies with camera traps, ranged across areas ten times bigger. Tracked zebras on annual migrations traveled five hundred kilometers round trip, one of the longest land migrations ever logged. Estuarine crocodiles in Australia, which were presumed to avoid ocean travel, swam over two hundred miles into the sea, traveling along ocean currents. Dragonflies migrated from the eastern United States to South America, flying hundreds of kilometers every day. Tiger sharks, assumed to be permanent residents of the coastal waters around Hawaii, turned out to travel thousands of kilometers out into the sea. Scientists' assumptions about their provincialism, a shark researcher from the Hawaii Institute of Marine Biology said, "were completely wrong."

In one particularly epic journey tracked by satellite, a wolf collared in Trieste, Italy, trotted over one thousand kilometers across frozen rivers, six-meter-deep snows, and 2,600-meter-high mountain passes, into Austria, traveling continuously for four months.

Wild species regularly roam beyond the borders that scientists have defined for them. Giraffes in Ethiopia spend most of their time outside the borders of the park that was specifically designed to protect them. Green turtles in the Chagos Archipelago swim beyond the boundaries of the marine protected area meant to contain them. Forty percent of the birds in the limestone caverns of Venezuela's El Guácharo National Park roost and forage outside its borders. Elephants that were thought to restrict their movements within Kenya wander across the border into Tanzania.

Their movements are not simple. The more extensively scientists track animals, the more complexity they discover. A three-year study in the Himalayas of tragopans, bright red pheasants that scientists understood to move uphill in summers and downhill in winter, revealed that they move both uphill and downhill in winter; some even migrate elsewhere altogether. One biologist, who mapped out badgers' underground burrows by checking on the animals' locations once every twenty-four hours, found more burrows in direct proportion to the frequency with which he checked. Even checking badgers' movements every three seconds, he discovered, wasn't enough to reveal the labyrinth of passages the badgers created to facilitate their movements. To accurately capture it, he'd have to sample ten times every second.

The physiological ease of animals' movements has been underestimated. Pythons from Southeast Asia deposited in Florida navigated back to the precise location of their release, a journey of over twenty kilometers of Florida swamp, straightforwardly and with speed, months later. A tracked leopard made it across three countries in southern Africa, successfully circumventing the towns, cities, and roadways that scientists had presumed would make the journey impossible. Bar-headed geese that flew over the Himalayas ascended from sea level to over six thousand meters not during the day, when tailwinds would help boost their flight, but at night, against headwinds. Researchers dubbed it the "most extreme migration on Earth."

The myth of a sedentary world had cast wild species as so limited in their capacity to move that their most far-reaching movements could only be mediated by humans. In fact, their ability to transport themselves in complex, sophisticated ways eclipses ours.

Movements that were once dismissed as robotically controlled by genes appear to be the result of dynamic interactions between individuals, each responding to subtle cues in the environment and from one another. Songbirds whose movements were thought to be controlled by genetic instructions to head south at a certain time coordinate the timing and direction of their movements according to subtle factors in the environment and cues from one another. Black warblers follow complex routes through the sky, taking advantage of a network of wind highways that swoop over the seas and continents. Baboons thought to robotically follow their leaders decide dynamically between different pathways. When two baboons move in different directions, their followers split the difference between their trajectories, plotting a course in between the two. Even the spiders thought to be passively carried around on winds actively climb to the tops of plants, where they attach their silken threads and stand on tiptoe, waiting for the breeze to gather them up.

"Humans trying to achieve this in the absence of modern technologies," noted the ecologist Iain Couzin, referring to the mass movements of insects, birds, and other animals, "would be unthinkable."

The study of animal movement, once relegated to the margins of biological research, has shifted toward its center. In 2006 a group of scientists gathered at the Israel Institute for Advanced Studies in Jerusalem to sketch the outlines of a new approach that would situate movement as one of the central features in the behavior of animals and the functioning of ecosystems. They called the new field "movement ecology." The following year Wikelski and his colleagues started Movebank, a public database where scientists can share their animal tracking data. Animal trackers add about a million data points every day.

~

On a February morning in 2018, a small knot of scientists stood on the edge of a snow-covered field in the steppes of Kazakhstan, clad in heavy black parkas and fluffy hats with ear flaps. The subzero wind chapped and reddened their exposed faces. They watched the horizon, where a Russian Soyuz rocket is about to blast into the dreary gray February sky. When the slim white cylinder lifts off, the fiery blaze behind it like a gash in the firmament, they stamped their feet, flung their arms around each other, and hollered with delight.

The rocket, speeding at seventeen thousand miles per hour toward the International Space Station, carried on board two hundred kilograms of antenna. A few days later two Russian cosmonauts on the station would embark on a five-hour space walk, for which they'd been training for months, during which they'd mount the antenna on the exterior of the station. And with that, they'd commence a new phase in human understanding of the scale and tempo of wild species' movements across the planet.

During every orbit, the antenna would scan the surface of the earth sixteen times, picking up data from thumbnail-size solar-powered tags ecologists across the globe had fitted on the backs of fish, the legs of birds, and behind the ears of mammals. The scientists would be able to control and reconfigure the tags whenever they liked. To begin with, they'd stream a continuous log of the animals' locations along with clues to their behavior embedded in data about their orientation, as well as factors such as the temperature, humidity, and pressure around them. The new satellite-based system is called the International Cooperation for Animal Research Using Space, or ICARUS. It's been described as a kind of "internet of animals," illuminating in real time the intricate web of animal tracks across a dynamic earth.

Given movement ecologists' insights into biological processes gleaned from tracking individual species, Wikelski predicted even deeper insights from tracking many species simultaneously. Seeing disconnected patches of sky hadn't allowed astronomers to understand the universe. That became possible only after they set up a network of telescopes to survey all of space at once. Movement ecologists hope to effect a similar revolution in understanding with ICARUS. "We see the whole network of animals around the world as one big information system," he said, "that is so far untapped."

Wikelski stood in the Kazakhstan snow wearing a black knitted hat with large white polka dots, his neck snug in a thick knitted scarf. After a round of bear hugs, he and the others went off to enjoy a celebratory shot of vodka.

∿

As new data about animal movements past and present pile up, ecologists have started to reevaluate their theories about the damage caused by border-crossing species on the move.

Invasion biologists who predicted ecological Armageddon caused by species on the move had underestimated the scale and tempo of wild movements, most of which were not disruptive. One analysis showed that only 10 percent of newly introduced species establish themselves in their new homes, and only 10 percent of those flourish in ways that can threaten already resident species. Condemning all newcomers as inevitably damaging blames them all for transgressions committed by 1 percent or less of their members.

When the Suez Canal artificially joined the Mediterranean with the Red Sea, which had been separated for millions of years, over 250 species moved from one side to the other. Their movement had led to the single extinction, according to scientific assessments a century later, of a sea star called *Asterina gibbosa*. The introduction of eighty marine species into the North Sea, and seventy species into the Baltic Sea, led to zero extinctions among the locals.

Because displacements don't happen at the scale that invasion biologists predicted, the arrival of newcomers increases biodiversity. In a paper that was rejected by *Nature* because of the way the public might misinterpret it, the Canadian ecologist Mark Vellend found that wild newcomers generally increase species richness on a local and regional level. In the continental United States, four hundred years of open borders to wild migrants has increased biodiversity by 18 percent.

To calculate the burden caused by wildlife on the move, invasion biologists had included not just the damage that newcomers cause but the cost of preemptively getting rid of them, too. They excluded the economic benefits of wild migrants. Including the benefits only of the introduced plants that contribute to the global food supply added $800 billion to the plus side of the equation.

When the botanist Ken Thompson compared the impact of successful introduced species to that of successful resident species, he found "in almost all respects they were the same." Even some of the most prominent newcomers failed to fulfill invasion biology's predictions. Zebra mussels cannot be blamed for the collapse of native clams, which face a number of challenges besides the newcomer's appetites. And besides disrupting local ecosystems, the mussel also contributes to them, by filtering water and providing food for fish and waterfowl. "If zebra mussels were native," Thompson noted, "there's every reason to expect they would be hailed as environmental heroes, rather than vilified as public enemy number one."

When Canadian researchers compared plots with and without purple loosestrife, they found that the plant neither reduced diversity nor displaced native species. There is "certainly no evidence that purple loosestrife 'kills wetlands' or 'creates biological deserts' as it is repeatedly reported," a 2010 review paper concluded. Their biggest crime, Thompson writes, is being conspicuously successful. Even that hasn't lasted: in places where they've been present a while, they tend to decline.

"The classifications of species as either 'native' or 'alien' is one of the organizing principles of conservation," a 2007 review paper noted, but "the validity of this dualism has increasingly been questioned." Species move around in ways that belie any simplistic categorization as "native" and "alien," Thompson says. In his book *Where Do Camels Belong?*, Thompson noted the case of the camel, depicted on animal maps as being "native" to the Middle East. But the camel family evolved and attained its greatest diversity in

North America, is presently most diverse in South America, and occurs in the wild only in Australia.

Thompson and other critics of invasion biology don't dismiss displacement of local species as an urgent problem. On remote islands, for example, introduced species could effect dramatic displacement of already resident species. But even in such places, it isn't only newcomers that disrupt and encroach on locals. Natives do, too.

~

In the jungles along Mauna Loa, efforts to rid a patch of forest of intrusive newcomers failed miserably. Rebecca Ostertag and Susan Cordell had tried for years, but to no avail. Even if the botanists removed every foreign weed and captured every alien seed that rained down from above, the newcomers would keep coming back. The invisible seeds and spores of the newcomers infest the ground all around. Keeping even one of the tiny plots free of foreign contamination required forty hours of hard labor a week. "It was absolutely way too much," Ostertag said.

The invasion of species on the move appeared unstoppable. Finally, Ostertag just gave up. "Getting back to an all-native system," she said, "is completely unrealistic."

But it wasn't just futile. It also seemed unnecessary. Ostertag and Cordell came to realize that the native species in Hawaii are not necessarily any more ecologically functional than other species. When Ostertag graphed the functional traits of Hawaii's native species on a chart, they all clumped together in a corner. The native ecosystem had been "dysharmonic," Ostertag says. It was like a picnic consisting entirely of potato salad. Whole functional groups were missing. There'd been no amphibians, no mammals, no reptiles, no ants, and no gingers among the plants. Because they all had to be able to survive under Hawaii's harsh conditions, the native species of Hawaii were a peculiar group. That's why the newcomers who followed thrived the way they did. It wasn't that they were rapacious aliens with sharp elbows. They filled ecological openings that the locals had left open.

A few days before I arrived in Hawaii, the fungus that had been decimating Hawaii's *'ōhi'a* trees was identified. Based on its behavior, most scientists had presumed the killer would be an interloper. But it turned out that the fungus could be found nowhere other than Hawaii. No one came out and admitted it, but the killer could only be called a "native."

Ostertag and Cordell devised a new experiment, taking into account how natives and aliens might live together. Instead of ridding the forest of newcomers, or letting them take over entirely, they aimed to rebuild a stretch of the forest into a mixed, diverse community comprising both newcomers and old-timers, natives and aliens. They chose which plants to grow and nurture based not on where they had come from or when, but on their traits and what they could contribute to the ecosystem. They spent three years setting up the experimental hybrid ecosystem. By the time I visited, the trees there had grown more than twenty feet tall, and the canopy was beginning to close. The rate of new arrivals had stalled as the forest floor shaded over, depriving new seedlings of light, a sign that the hybrid forest was maturing into a self-sustaining one.

As we made our way to the field site, I asked them what the native forest might have looked like. Neither Ostertag nor Cordell claims anymore that natives are any better ecologically than newcomers. But their affection for the old-timers is still palpable.

Cordell, wearing gold hoop earrings and her hair pulled back by sunglasses perched on her head, describes how she imagines these jungles before the newcomers arrived. The ʻōhiʻa and other native trees would have dominated the overstory of the primeval Hawaiian forest, she says, lianas and vines dripping from their canopies to the forest floor. The ground itself would have been blanketed by lush stands of tree ferns, she tells me.

I look down at the ferns at our feet. They are scattered across the forest floor, sprouting alone and in little clumps. They're still here, I say to Cordell.

"Yeah," she says, drawing out the word. "But these aren't native unfortunately."

Is there something wrong with them, I ask, besides that?

"I don't know," she says. "That's a hard question. I mean, when I started in conservation, I would have said everything nonnative is bad. But I don't think that way anymore. This project has really turned my world." She pauses. "I mean, this is our life! This is where we are in the world! And we're scientists. And isn't this interesting to study?"

*It's possible*, I think, *that she's trying to convince herself.*

She looks down at the fern. "And a lot of them are pretty, you know."

~

Funza, Colombia, is a small town in the Andes, about eight thousand feet above sea level. What's now a high plain lay, during the Pleistocene era, at the bottom of a lake. In 1989 the geologist Lucas Lourens and his team positioned a Portadrill truck just outside the village, drilling a narrow hole nearly six hundred meters down, until they hit the bedrock.

The sedimentary core they brought up represented a record of the species that had lived in the area over millions of years. The remains of animals had long disappeared, but the pollen shed from the plants, trees, herbs, and shrubs that had grown in the area had settled in those sedimentary layers, one atop the other, sinking deeper over time.

I learned about what Lourens and his team found thanks to a casual mention of their 2013 paper in Ken Thompson's book about the folly of splitting wild creatures into natives and aliens. As far as I know, its findings had never reached a broad audience, the way alarmist stories about foreign plants and animals often did. There'd been no magazine stories or radio episodes about Lourens's study. But the glimpse into biological history that it offered struck me as deeply moving.

The pollen revealed a continuous process of climatic change and migration. As the landscape transformed, new species migrated in and out. There was pollen from swamp forest trees, button-weed, cypress-like shrubs, heathers, and medicinal herbs. When the Panamanian isthmus rose up, unleashing migrant flows between the North and South American continents, pollen from oak trees arrived.

The types of species that Lourens and his colleagues found, and the combinations in which they found them, never repeated. At each moment in time, the species that inhabited this slice of land in Colombia were completely new, living in mixed communities that would have been unrecognizable to denizens that had been there before and that would come afterward. Each layer of the sedimentary core represented a single "frozen moment," they wrote, in a "long and dynamic process of almost continuous reorganization."

Climatic regimes came and went. Geological eruptions made their slow-motion advances and recessions. Sea levels rose and fell. Monkeys crossed oceans. Ferns colonized Hawaii. Koa trees sired their progeny on Réunion Island. *Homo migratio* left Asia and canoed into the Pacific, guided by the stars.

With each change, new opportunities for species on the move opened up. As those opportunities arrived, the migrants came.

That's because nature transgresses borders all the time. And with good reason.

9

# THE MIGRANT FORMULA

The sun has nearly set on a moody October evening when we see the bear cross the border. The old logging road we walk on, barred to traffic by a metal gate, seems to have been graveled at one time. Now a soft carpet of moss, shading from light sage to deep puce in the twilight, tapestries its surface, punctuated with spiky clumps of pale gold grasses. The trees of the northern Vermont forest that line the narrow road stand gray and shorn, having long lost their leaves in preparation for winter. But not the beech trees, which are migrants from the tropics and so retain a vestigial clutch on their canopies. Their marigold leaves tremble delicately in the breeze, scattering light over the gray forest. Claw marks scar the tree trunks, evidence of the bears that come from hundreds of miles away, drawn by the migrant trees' rich fatty nuts that nourish them for their long winter sleeps.

The road descends gently into a wooded valley, surrounded by low, old hills. Somewhere below us, an invisible line separates the woods into the Green Mountains of Vermont on one side and Quebec's Sutton Mountains on the other. Even as we approach it, there is no telling. The forest stretches over the rolling hills, unbroken.

But because of this invisible line, the forest here is booby-trapped with hidden, battery-powered cameras, painted gray and brown to match the woods. Sometimes the bears destroy them, especially if the border agents who set them up have stopped for a breakfast sandwich on their way to the woods. The bears find the scent left behind on the devices by the agents' sausage-grease-moistened fingers irresistible. Those that stay intact are monitored by Border Patrol officers stationed in a lonely, overly chilled facility a few miles to the south. Their intent is to detect the illicit movements of cars and drug smugglers. Mostly their cameras capture the movement of wildlife, who trip their shutters some two hundred times a day.

We've been hiking for about an hour when a large white Border Patrol van emerges out of the shadows. Two baby-faced agents, their bodies packed into stiff uniforms, politely but sternly inquire about our itinerary. The tension in their bodies almost instantly dissolves when they understand we are just observers and not migrants ourselves. They happily chat with us about the images of wildlife on the move shot by their hidden cameras. They download their favorites to keep in their own personal collections, they tell us. I will see them, later. The portraits—besides those of me and my guide, the wildlife tracker Jeff Parsons, looking rather suspicious in this context—are beautiful and haunting: the large brown eyes of a passing deer, the matted backside of a bear, the lithe torso of a bobcat fill their frames. They even caught one of the rarely seen velvet-coated beauty, the lynx.

The tracks of the animals moving across the land here run parallel to those in the sea and the sky, uncaptured by Border Patrol cameras. In the waters a few hundred miles to the east, great white sharks patrol the American coast, gliding from U.S. into Canadian waters and back again. Right whales carve a long watery trail from their calving grounds around the Florida coast to their feeding grounds in Nova Scotia's Bay of Fundy. Leatherback turtles chase clouds of jellyfish from their breeding grounds in the tropics deep into northern waters. In the sky, monarch butterflies heading south cross paths with ospreys and black-throated blue warblers en route from their winter homes on the island of Hispaniola to their summer homes in the forests of southern Canada.

Finally, the Border Patrol agents leave, and we continue our hike. At the bottom of a hill, we encounter an overgrown three-foot-wide path cut into the forests, running from east to west on either side of the logging road. It's the international border between the United States and Canada. The two powerful states' attempt to carve the landscape into separate pieces controlled by their authority has been made manifest in this narrow, dark passage through the forest. We stand there for a while, as the sun sinks below the hills, then head back up to find our parked cars. At the crest of the hill, we turn around to catch another glimpse of the border. In the twilight, I can just barely make out the small black bear ambling across it.

~

Biologists such as Elton and others who'd dismissed migrants as suicidal zombies and mindless invaders never really examined migrants' behavior itself nor considered in any depth how it might have evolved. What drives creatures to move into new territories, away from where they were born? Leaving behind the known comforts of a home habitat, migrants strike out into the unknown. They forsake the help of kin who stay behind. In return, they may not find anywhere suitable to live at all.

And yet they do it anyway.

Baleen whales migrate thousands of miles from their rich feeding grounds in the far north to the warm waters of the tropics. Zooplankton migrate vertically between the depths and the surface in sync with the fluctuating light. Forests move over thousands of years with the advance and retreat of glaciers. In jungle-dripped Hawaii, tiny goby fish migrate from the open Pacific Ocean back to their birthplaces at the tops of waterfalls. The journey requires swimming against ocean currents into fresh waters and climbing up cliffs. They use suckers on the undersides of their bodies to do it.

In humans, the origin and ecological role of migration continue to be shrouded in controversy and contention. But biologists have forwarded a clear idea about its provenance in animals.

Migration experts such as Hugh Dingle say that migration most likely evolved as an adaptive response to environmental change. Migratory behavior is more common among species that depend on resources exposed to environmental variations than in those whose livelihoods are more buffered from environmental change. Arthropods that live in temporary habitats such as shallow pools and seasonal ponds, for example, are more likely to migrate than those that live in relatively stable environments such as forests and salt marshes. Species that live in places with erratic rainfall or that feed on patchily distributed resources such as fruit and flowers are more likely to migrate than those that live in relatively stable places such as alpine tundra or deep lakes, where a disproportionate number of insect species don't even have wings. Species that live on the edges of forests or in their canopies are more likely to migrate than those that live in their interiors. Bird species that feed on fruit, which is available only seasonally, tend to migrate more than bird species that feed on insects in the interior of forests, which are not. Bats that roost in trees, where they're more exposed to the cold and the rain, migrate more than bats that roost in caves, which protect them from the elements.

Even within species, individuals that live in habitats exposed to change migrate more than those less exposed. The migratory behavior of white-tailed deer, for example, correlates with the size of their forest patches: deer that live in small patches more exposed to changing conditions migrate more frequently than deer that live in large patches.

For creatures that live in habitats subject to environmental change, survival rests on one of two strategies: either go dormant and wait for the altered conditions to recede or migrate. Again and again those creatures capable of movement have opted for migration, despite the costs. Dingle distilled a formula that predicts the emergence of migratory behavior. It lies in the ratio between the time it takes to reproduce a new generation and the stability of the environment. If that ratio is less than one—if, say, it takes a couple of years to reproduce the next generation but the habitat, say a vernal pond, lasts only for a season—migration is likely to emerge.

And so as the northern hemisphere tilts away from the sun, lengthening shadows and shortening days, creatures of all kinds prepare to move. Physiological changes transpire inside their bodies, hormones spiking and nervous systems mobilizing. Sap-sucking rosy apple aphids birth special forms with wings. Salmon on the verge of migration experience spikes of hormones such as prolactin and cortisol. Baby eels metamorphose into transparent forms that prefer fresh water to salt water. In preparation for their travels, migratory birds and insects build up fat stores that can comprise more than 50 percent of their body mass, and plants produce tough coats on their seeds. The proportion of fat they deposit in their seeds correlates to the distance the seeds are likely to travel.

As the time to leave approaches, a restlessness sets in. Migratory birds trapped in cages will flutter repeatedly to one side of a cage, jumping off their perches and crashing to the side. Which side depends on the direction it faces: whichever matches that of their migratory path. Scientists named their agitation *Zugunruhe*, German for "migratory restlessness." It's hormonal. Remove the gonads of sparrows in the spring, and they'll be less restless; castrate a migratory bird, and it will still migrate, but in a different direction.

Migratory journeys are not simple extensions of everyday movements, like flying from one tree to another or moving from one cave to the next. In birds, you can tell as soon as they leave, from their rate of climb and the altitude they achieve, that their migratory flights are something different. While en route, their behavior and bodily functions are fundamentally altered. Unlike during ordinary movements, during migrations their bodies halt their own growth and development. They ignore stimuli they'd ordinarily respond to, passing by appealing foods and breeding spots.

But while migration is driven by physiological changes, it is not necessarily the result of some fixed itinerary instilled into the bones or some blueprint encoded in genes, propelling creatures in fixed directions at standardized times. The physiological states required for migration can be flexible and dynamic, too. The muscles that power the wings of migratory aphids, for example, start to break down after migration, the proteins diverted to reproduction instead. Animals' sensitivity to the quivers and trembles of our pulsating planet—not predetermined internal programs— drives their behavior and movement.

Wild animals' sensitivity to environmental perturbances is the stuff of legend. Anecdotal stories of animals apparently sensing impending environmental disruptions hours or days before they're detectable to humans stretch back to ancient times. Pliny the Elder described birds' restlessness before earthquakes. Geese detected the arrival of an invading army of Celts in 387 B.C.E. Rome before the sleeping residents did, their quacking alerting them to the impending onslaught. In 1975 snakes outside the city of Haicheng in China emerged from their hidden shelters and froze to death in the winter's cold, in advance of a 7.3 magnitude quake. In 2004 elephants in Sri Lanka fled inland hours before a tsunami reached the shore, saving the lives of those who instinctively followed them away from the wall of water.

The signal that the goats living on the slopes of Mount Etna in Sicily detected on a winter's day in 2012 remains obscure. So does the perceptual mechanism by which they'd detected the signal. But whatever it was and however they'd sensed it, they did so faster and with more sensitivity than any machine devised by humans could have. For nine months, animal trackers who had fitted the goats with transmitter collars had been watching monitors streaming data recording the goats' wanderings, as they ate, slept, and roamed across the volcano's slopes. They were watching at the moment when the goats' movement patterns changed dramatically. The event that triggered their burst of motion became clear six hours later, when the volcano erupted, spewing lava out of its crater for over twelve hours and shooting ash seven kilometers into the air.

~

Migration's ecological function extends beyond the survival of the migrant itself. Wild migrants build the botanical scaffolding of entire ecosystems. They spread pollen and seeds, shaping where plants live and in which proportions, and ensuring that seedlings can reach open habitat rather than withering in the shade cast by their parents. The transport their movements provide is so critical to plant survival that many plants have evolved ingenious methods of enticing animals into ferrying their seeds around. They coat their seeds in sticky mucus or produce hooks, spines, and barbs around themselves to hitchhike rides on passing mammals by clinging to their fur, as anyone whose dog has been covered in burrs knows. They produce seeds with fatty bits to attract ants, which carry the seeds and helpfully bury them underground. They produce fleshy, fragrant fruits around their seeds, to entice birds to ingest the fruit and scatter the seeds across their flyways in drips of poo.

Botanists say that the survival of over 90 percent of the trees in rain forests hinges on birds and other animals on the move scattering their seeds. GPS tracking studies have shown that even species reviled as parasites shower the landscape with seeds. The nineteenth-century naturalist Alexander von Humboldt dismissed cave-dwelling *guácharos*, or oilbirds, as parasites, noting their habit of fruit-eating inside their dark lairs, which condemned seeds to oblivion. In fact, they spend their nights flying across the forests of Venezuela, dispersing seeds along the way. They're "probably responsible for much of the diversity of the rain forest," the ornithologist Martin Wikelski says.

Wild migrants ferry genes into isolated populations, introducing life-saving genetic diversity. In small isolated populations, genes with once-diluted effects in the population, such as those that code for life-threatening defects or increase vulnerability to disease, become concentrated. As increasingly related couples mate, genetic homogeneity sets in, reducing the ability of their population to withstand disease and disaster. Ecologists saw the dramatic effects in a population of wolves on an island in Lake Michigan called Isle Royale. The wolves all descended from a single breeding pair that had arrived during a particularly frigid winter in 1949, when the channel between the island and the coast froze over, allowing the two to pad over on foot. Since then, the population they founded had been marooned. They became increasingly inbred. By 2012, 58 percent of the wolves of Isle Royale had congenital spine deformities, compared to just 1 percent in wolf populations elsewhere. Many had eye abnormalities, with one eye appearing opaque and possibly unseeing. A female wolf died in her den, with seven dead wolf pups in her womb, a single squealing live pup beside her. Ecologists had never seen anything like it before.

The colony's sole hope: migrants. In 1997 a single male wolf made his way to the island. The migrant's jolt of genetic rejuvenation single-handedly transformed the ecosystem. Within a generation, the migrant's genes lurked in 56 percent of the wolf population. The number of wolves on the island rose. The number of moose, which the wolves hunted, fell. The moose-trampled forests recovered. The lone migrant "saved the population for another ten to fifteen years," the ecologist Rolf Peterson said.

While isolated habitats deprived of the largesse of animals on the move suffer, those that facilitate animal movements flourish, as a handful of large-scale experiments in forests have established. In one, ecologists cleared several fifty-hectare patches of mature pine forest along the Savannah River in South Carolina, removing trees and burning vegetation to the ground. They arranged the patches with one in the center and the others arrayed around it. The dense forest that enclosed each patch formed a kind of border around them, disconnecting them from each other. Then they pierced those borders by building a corridor, clearing a single, twenty-five-meter path between the central patch and one of the peripheral patches. They then tracked how plants, insects, and pollen spread from the central patch to the connected and disconnected peripheral patches. In the central patch, they marked butterflies. They doused fruiting shrubs that birds feed on with fluorescent powder. They planted male holly bushes needed to pollinate the female holly bushes planted in the peripheral patches. Then they visited the connected and disconnected peripheral patches, counting the number of marked butterflies, the amount of bird poo with fluorescent-colored seeds, and the number of flowers on female holly bushes.

Opening borders and clearing a path for butterflies, seeds, and pollen allowed them to spread at least twice as fast to the connected patch as to the disconnected ones. By the end of the study, the connected patch was blanketed with flowers, fruit, and butterflies.

~

Migrants saved the checkerspots.

At the southern end of the Santa Clara valley, just a few dozen miles from where Paul Ehrlich studied checkerspot butterflies, lies an undeveloped expanse of hills called Coyote Ridge. Coyote Ridge is prime butterfly habitat: thousands of acres of butterfly-friendly patches are home to varying mixes of sun, shade, and soils, plentiful wildflowers, and even fuzzy light brown calves and their doe-eyed mothers munching on the grasses that compete with the butterflies' preferred host plants. If conditions deteriorate on one hillside, butterflies that live here can easily move to another. No miles of intervening highways and shopping malls block their passage. When conditions are good, they can sharpen their genetic adaptations to their locales, adding to the overall genetic diversity of their populations.

Ehrlich and his students hadn't known about the butterflies living here. Butterfly collectors, on whose historical records butterfly biologists long relied, never ventured into these lovely hills. Once they'd found sufficient specimens to stock their collections, they'd stopped looking. But the checkerspot butterflies that lived in Coyote Ridge—and their migratory connections to other butterfly patches—provided a lifeline to the colonies that Ehrlich and Parmesan studied.

Over the course of thirty-five years, checkerspot butterflies reappeared in about 13 percent of the patches that Parmesan and her colleagues monitored. "It was like, 'Wow! Cool!'" Parmesan says. They reappeared on the ranch in Arizona where their colony had collapsed entirely seven years earlier. Somehow butterfly pioneers had emerged. And they'd overcome miles of formidable, uninhabitable territory, located suitable new habitat, and established new colonies.

Their survival seemed "something of a paradox," as one butterfly biologist noted in an *Ecology* paper. Ehrlich and others had dismissed the movements of butterflies and other "small, feeble insects" as accidental and ecologically meaningless, as a prominent entomologist who studied migration wrote. Ehrlich figured that just 3 percent of his checkerspot butterflies would move even 50 to 100 meters between patches. But in at least one study, biologists recaptured checkerspot butterflies six miles away from where they'd been released.

Just what might have triggered their movements remains obscure. Perhaps it was an inadvertent outcome of abundance. By triggering population booms, especially lush conditions might have increased the likelihood of either a rare wanderlust-struck butterfly emerging, or a less adventurous one being swept up in a breeze and successfully deposited in a new habitat. But the butterflies' migratory urge might also have been a response to scarcity. Suggestively, caterpillars that experience malnutrition increase their investment in flight muscles, compared to well-fed caterpillars. And signals of diminishing resources trigger the emergence of winged forms in ants and termites, equipped to travel.

The morning I visit Coyote Ridge, the sun struggles to burn through a thin blanket of clouds, and a steady damp breeze whips the pages of my notebook, sending strands of hair flying into my face. Despite the weather, it still looks like a scene out of *The Sound of Music*. Low grasses and wildflowers blanket the broad hilltop meadows, one after the other, as far as the eye can see, interrupted by scattered outcrops of serpentine rocks covered in bright orange lichen. The hum of the highway that leads to the sprawl of Silicon Valley is barely audible. In the distance, the Diablo mountain range looms, distantly dotted with herds of elk and deer.

Little flocks of checkerspot butterflies flit around my feet. They're everywhere. Migrants dart among them, connecting them to distant habitats with fragile, silken threads.

∿

Scientists have illuminated the origins and ecological functions of migration in animals by amassing indirect data. They can't ask a wolf padding through snow-capped mountains where it's going and why. But the migratory longings of humans can be investigated directly.

Whenever I meet people on the move, I ask them the same question. The men and women camping in an abandoned stadium outside Athens, suffering epidemics of scabies and washing their clothes and children in filthy public bathrooms while standing in an inch of stagnant water, who'd arrived from Pakistan. The woman who'd landed in Baltimore with her son, after leaving her three-year-old daughter and her parents on a farm in Eritrea. My father, who still yearns for the choked tenements of Mumbai where he'd grown up, more than fifty years after he left. My father-in-law, who left postwar Britain with nary a look back. I asked them all: Why did you leave?

"When you first called, I had the anxiety," one migrant tells me, when we meet in a cramped ground-floor office in a small brick building outside Boston. "I did not want to call you back. There is no way for me to come," he says he remembers thinking. He'd come anyway, driving forty-five minutes to meet me.

But the man, who'd migrated to Boston from Haiti, says he cannot tell me what I came to hear. The documents that allowed him to enter the United States are now in question. He won't give me much more than impressions about his life before and after he migrated by way of explanation for his journey.

The boy from Kabul teared up, when I asked him. He wanted to study electrical engineering, he told me, and had planned to apply to study it at a specialized school in Kabul. "Nobody wants to leave their country," he tells me, but "everyone is in danger there. You walk, and there's bombs going off." Fearing he'd be recruited by the Taliban, his family sold everything they had to send him on a long trek not just out of the Taliban's reach but all the way to Europe, on foot, accompanied by some distant relatives. He was in eighth grade when he set off, leaving his parents and older sister behind. But why him, why then, why not others? I wanted to know. He could not tell me.

One man who had left a farm in Haiti for a chilly flat on the outskirts of Montreal smiled nervously when I asked him. "I know someone beat me," he says to me. "They make me die." He cannot answer any follow-up questions. I don't know what to think. But I know that his future depends on how he answers this one particular question of why. If he can convince the immigration authorities that he left for what they consider the right reason, he may be able to stay. If the authorities feel that his reasons are not correct, he'll be forced to leave. Opportunistic immigration "consultants" gouge migrants like him for hundreds of dollars that they don't have to help them mold their messy stories into tales that will allure the authorities, whose demands for reasons behind migrants' movements change with the political winds.

We can ask migrants why they move, but it's not necessarily possible for them to answer, at least not in the direct and simple way we'd like. The question assumes that human migration can be explained by some singular reason. That assumption shapes how we talk about people on the move. We describe them as "economic migrants" or "political refugees." Some characterize them by their suspicions about their legal status as "aliens" or "illegals." We define them by the directionality of their movement over international borders, as "immigrants" or "emigrants," submerging their equally if not more complex and lengthy movements within borders.

But all we really know is that they are people on the move.

~

Among migrants, *Homo sapiens* is king. And yet we have little consensus on why we move around the way we do. The findings of continuous migrations throughout our deep past have upended the idea that we moved only once in the past, attracted by empty lands, but have left the central question intact: Why? Why venture into the oxygen-starved Tibetan plateau or set off on outrigger canoes into the waves of the Pacific? Why leave the comforting certainties of life in Africa, where food and water and other resources abound to this day?

While the ecological role that wild migrants play is increasingly well documented, the motives and impact of human migration remain shadowy and ill-defined. Many popular theories locate the origins of human migration in nonmigratory behaviors, as if migration were essentially an accident, a by-product in our quest for other goals. The archaeologist J. Desmond Clark, for example, theorized that our first migrations began simply because we followed wild animals that moved. Our early ancestors hunted herds of wildebeest and antelope, he pointed out, and they moved seasonally over long distances. We followed, spears in hand, bellies empty, and in the process were drawn steadily farther and farther afield, turning us into accidental migrants.

Indeed, human movements continue to track those of wild creatures in modern times. In the seventeenth century, people from France migrated to North America in search of fur-bearing animals, whose pelts they'd use in their felted hats and the like, establishing the colony of New France in Canada. In the late eighteenth century, people from the Azores Islands migrated to New England, following the whales they hunted, establishing still-extant Portuguese communities in Massachusetts. We moved in sync with the animals because our livelihoods depended on their fur and flesh.

Most of our livelihoods no longer depend directly on animals and their movements, but all the same we still move in order to secure the economic sustenance they once provided us. Most every migrant could accurately describe their movements as motivated by the desire for work and economic security: for all their travails in their home countries, the man from Haiti hoped to become a nurse, the boy from Kabul an engineer. And their labor is one of their most consequential impacts, adding billions of dollars to the economies of the countries they enter. Because so many send money back to their relatives and friends left behind, migrants' labor adds billions of dollars to the countries they left as well. International migrants send more than $500 billion to their home countries every year, a flow of money that steadily redistributes wealth across borders. For some countries, these so-called remittances from migrants living overseas form sizable proportions of their GDP. According to data from the World Bank, remittances account for around 20 percent of the GDP of Lebanon, Nepal, and Moldova.

Still, our migration patterns cannot be defined solely as a product of the search for jobs. Economists have tried. In one formula, neoclassical economists calculated the likelihood of migration based on the difference between wages here and wages there: $ER(O) ffl [PI (t)P2(t)YOt) - P3(t)Yo(t)]ertd - C(o)$. It looks more like the way to calculate the rate of nuclear fission than the probability of a messy human endeavor such as migration. And it doesn't actually work.

~

Other popular theories about the origins of human migration suppose that some systematic change in the climate triggered our first movements out of Africa. Some, presuming migration to be an act of desperation, imagine it must have been a sudden, catastrophic event. The eruption of Mount Toba in Indonesia 74,000 years ago, for example, blanketed the skies with ash and depressed global temperatures for millennia. Perhaps that long volcanic winter "precipitated a desperate search for new food and land," as Siddhartha Mukherjee put it in his popular history of genetics.

The migratory response to future environmental change is similarly cast as a last-ditch one, forced by catastrophe. In white papers and articles, national security and foreign policy experts issue predictions about how the disruptions and dislocations of climate change will affect migration. Food and water shortages will lead to instability, which will force migrants into motion, leading to more instability. Catastrophic floods and expanding deserts will force whole communities to pick up and leave. Rising seas will inundate millions of homes, forcing their residents to flee. By translating each "unit" of climate change into a proportional additional unit of migration, as the geographer Robert McLeman put it, experts such as the environmentalist Norman Myers estimated that by the mid-twenty-first century, climate change will create an army of 200 million environmental refugees, who will scour the planet. Migration will be "one of the gravest effects of climate change," the Intergovernmental Panel on Climate Change noted, "one of the most dramatic consequences of global warming." Such climate-driven migrations could even lead to civilizational collapse. According to their assessments, it has happened before.

But perhaps migration takes hold during periods of opportunity, not crisis. It's possible that our restless ancestors, rather than reluctantly escaping from bad conditions, capitalized on good ones. The earth's orbit wobbles on time scales of tens of thousands of years, the rotations switching from elliptical courses to circular ones. These orbital shifts change the angle and strength at which

# Human settlement patterns and future climate change

Human populations around the world will face an array of future climatic changes. How these changes will shape the coming era of migration remains to be seen.

Relative population densities

**Areas likely to be affected by:**

Sea-level rise, surges, and floods

Hurricanes and storms

Droughts and desertification

Melting permafrost

Major climate crises

WESTERN ALASKA

ARCTIC OCEAN

RUSSIAN WESTERN ARCTIC

Canada

United States

EAST COAST

Louisiana

FLORIDA

Mexico

VERA CRUZ

CARRIBEAN ISLANDS

Costa Rica

Colombia

Antigua

Guyana

PACIFIC OCEAN

ATLANTIC OCEAN

AMAZONIA MARGINS AND NORDESTE

Brazil

Peru

Argentina

RIO DE LA PLATA

Netherlands

PO DELTA

NORTH AFRICA

CENTRAL ASIA

Russia

Egypt

NILE DELTA

SAHELIAN COUNTRIES

Gambia

Nigeria

NIGER DELTA

Ivory Coast

EASTERN AFRICA

Mozambique

Seychelles

Mauritius

Réunion (France)

INDIAN OCEAN

Maldives

India

CENTRAL INDIA

Bangladesh

CENTRAL CHINA

China

TIANJIN

YINGKU

SHANGHAI

PEARL RIVER DELTA

Vietnam

MEKONG DELTA

Philippines

PACIFIC ISLANDS

PACIFIC OCEAN

Australia

Sources: Data compiled from Intergovernmental Panel on Climate Change; World Meteorological Organization; Dina Ionesco, François Gemenne, and Daria Mokhnacheva, The Atlas of Environmental Migration (London and New York: Routledge), 2017; United Nations Environment Programme

the sun's rays hit the planet and so, over time, alter the planet's climate. Such climate swings might have facilitated human migration, by turning the impenetrable deserts of North Africa, for example, into habitable savannah-like green corridors across which humans might have moved, like butterflies and clouds of pollen across the forests along the Savannah River. Suggestively, computer modelers from the University of Hawaii have found that orbital climate changes correspond with the pulses of human movements out of Africa.

~

Our fears and confusion about why people move permeate the laws we've passed regulating whether we're allowed to relocate and under which conditions. Migrants' job-seeking, despite its powerful economic impact both on the societies they enter and on those they leave behind, is only sometimes and in certain places considered a legitimate reason to be granted permission to move across an international border. In places such as the United States, conflicting policies cater both to employers who benefit from free movement of labor and to workers who feel threatened by it. The contradictory result—a flow of newcomers who are simultaneously allowed entry but are stigmatized for doing so—emerges from the political tension.

Authorities' positions on the legitimacy of moving to escape hardship are equally mercurial. Melting glaciers and rising sea levels have already made scores of towns and villages on low-lying islands such as Kiribati in the Pacific, Isle de Jean Charles in the Gulf of Mexico, and Shishmaref off the coast of Alaska uninhabitable. Even more people have been compelled into motion because their fields have dried up or their crops have failed, one of the many outcomes of climate change that scientists have long predicted. Most would not describe themselves as being displaced by climate change, though they arguably are. Out of all the countries in the world, only New Zealand has considered the idea of letting them cross international boundaries for that reason. In the United States, people fleeing natural disasters and armed conflicts can enter under the country's "temporary protective status" program, but only for limited periods of time, regardless of how permanently their homes and communities are damaged.

Instead, the 144 countries that signed on to the 1951 Refugee Convention offer refuge to only those migrants fleeing a certain kind of abuse and oppression. Refugees, by its definition, are those who flee state persecution of members of their race, or their religious or social group, that is, the kind of abuse doled out by the Nazis, whose crimes motivated the formulation of the convention in the first place. Those who move across borders to escape oppression and abuse doled out through other means—the oppression of poverty or environmental degradation, say, or the abuse of a failed state that refuses to police their communities or educate their kids—do not qualify, although they fit perfectly into our colloquial understandings of the term "refugee."

Some countries have signed treaties with each other so that they can refuse entry even to those fleeing Nazi-style persecution, if they'd passed through any other country where they might have been able to apply for refuge en route. According to these "Safe Third Country" agreements, if migrants' journeys to, say, Canada wend through the United States, or to the United Kingdom through, say, Greece, authorities can force them to turn back, even if they are fleeing the kind of abuse sanctioned by international law. In the summer of 2019, the Trump administration attempted to force poor and unstable countries such as Guatemala to sign Safe Third Country agreements, too, threatening catastrophic tariffs if they didn't comply. According to the logic of such agreements, fleeing state persecution is a legitimate reason for migrants to be granted entry only if they are truly desperate, and if they've taken their very first opportunity to apply. If they didn't, it wasn't. They'd be sent back.

The migrants I speak to know all this. I don't think they're not telling me the truth. But their stories must run through filters, like murky water through sand.

When I was small, like a lot of children I had a habit of asking overly broad "why" questions, like why airplanes fly, then refusing to be satisfied with whatever answer adults provided, continually demanding further explication. In one oft-told tale in my family, my uncle abruptly ended one of our long series of back-and-forths by marching me outside and pointing to the sky. By then I had likely asked him why about a dozen times in a row. "You see the sky?" he asked me, exhausted. "It's so high. And you can't fly. That's why."

Some phenomena, in other words, are not necessarily amenable to simple explanation. In the case of migration, I've come to realize that asking why people move reveals more about us and our expectations and fears than it does about migrants or migration. The idea that there should be a single explanation for migration is "rooted in a sedentarist notion," the geographer Richard Black suggests, according to which "migration is seen as a problem or exception from the norm, which needs explanation."

~

While confusion reigns about why we move and which reasons, if any, should be considered legally acceptable, telling evidence suggests that migration is encoded in our bodies, just as it is in wild species.

While we may not be able to detect volcanic eruption like the goats of Mount Etna, human bodies are sensitive and responsive to environmental change, too. Our relatively modest number of genes—about the same number as a nematode worm—does not translate into a narrow range of functional, developmental, and morphological differences among us, because our genes function in dynamic interaction with the environment. They're like letters in an alphabet, capable of expressing a wide variety of meanings, depending on their pattern and context.

Our bodies have equipped us with a range of options, producing results that suit a diversity of conditions. Over 180 different genes influence our height. At least eight different gene variants influence the color of our skin, each instructing skin cells to produce variable quantities of pigment. In notable distinction to the principles of eugenics, all those gene variants originated in Africa and are present in the genomes of both dark-skinned and light-skinned people today.

The frequency of genes, the presence or absence of other genes, and the environmental microconditions around our genes can dramatically change the way our genes are expressed. Temperature, for example, can change the degree to which a gene is expressed or whether it's expressed at all. In fruit flies, certain genes are expressed at specific temperatures; at others, they're not. The color of light that falls on caterpillars will change the way their genes for wing colors are expressed. If reared under red light, they'll be intensely colored; under green light, dusky; under blue light, pale. Depending on the population density around them, desert locusts will develop into sedentary forms or migratory ones. Depending on whether they detect chemical traces of predators around them, semitransparent crustaceans called *Daphnia* will give birth to differently formed offspring, either with or without defensive helmet-like structures.

The environment around us shapes how our bodies develop, too. Inside our mother's wombs, we flail and toss, the pattern of our movements etching on our hands the unique furrows and ridges of our fingerprints. Meanwhile signals from the external environment seep through our mother's bodies into our own, whether it's chemicals in the air she breathes or the kind and quantity of the local foods she consumes. Our bodies respond to those signals, changing the way our genes instruct our cells to function and shifting the course of our development.

One of the mechanisms by which this occurs involves a process called methylation. Genes have little groups of methyl molecules around them that act like switches, turning genes off and on. This in turn can affect whether other genes get turned off or on, triggering a cascade of interactions. Cues from the external environment, such as our mothers' experiences of famine or their ingestion of pollutants, shape the process.

People born to women who'd been pregnant during a short-lived famine in the Netherlands during World War II, for example, have different methylation patterns in their genes from those of their same-sex siblings born before or after the famine. Their bodies absorbed signals of famine sent via their mothers, and they transformed as a result. Researchers found that people who experienced the "Dutch hunger winter" in utero have increased levels of triglycerides and low-density lipoprotein cholesterol in their blood, suffer higher rates of diabetes and schizophrenia, and have a 10 percent higher risk of mortality than people born before or after the famine.

Even after we're born, environmental conditions shape the development of our bodies. For example, at birth, we all have the same number of sweat glands. But the ambient temperature we experience during the first three years of life dictate how many of these glands will become functional, altering our capacity to withstand heat for the rest of our lives. If the weather is sultry during our first few years, we'll have more functional sweat glands and be better equipped to withstand heat; if not, we won't.

When people fanned out of Africa and into novel environments with different climates, foods, and pathogens, our bodies adapted in response. Different genetic variants spread to help us survive the peculiar microbes in the habitats we entered. People who encountered malaria-carrying mosquitoes adapted with genetic variants that protected them from malaria's appetites. People who lived around the Ganges River delta, where cholera lurked, evolved adaptations that reduced the risk of dying from the disease. People from that part of the world have the lowest rates of blood type O, which increases cholera's deadly effect.

When the weak sunlight of northern climes threatened humans with vitamin D deficiencies, narrowing women's birth canals and ratcheting up the risk that both they and their babies would perish during childbirth, gene variants that increased their ability to absorb vitamin D from the sun's rays proliferated. That environmental adaptation can be seen in the paler skin tones common among people who live in Europe, North Asia, and elsewhere.

Those who moved into cold regions developed higher metabolic rates and stockier bodies that reduced heat loss. People from circumpolar North America and Siberia have higher metabolic rates than other peoples: inland Inuit people's metabolic rates are up to 19 percent higher than those of non-Inuits, to this day. Genes that helped people digest meat likely spread in people who relied on the flesh of animals for sustenance. Genes that rapidly converted plant lipids spread in those who specialized in vegan diets, such as my own ancestors in India. Genes that helped digest lactose spread in those who relied on milk through adulthood. White nationalists consider their ability to digest milk a point of pride, holding events in which they ostentatiously chug gallons of the stuff, but in fact such genes are present in a range of peoples besides dairy-farming northern Europeans, including cattle-herding peoples in Sudan and the camel-herding nomadic peoples in the Middle East and North Africa known as the Bedouin.

When life in the oxygen-poor heights of the Tibetan plateau felled pregnant women with preeclampsia, genes that allowed us to withstand the perils of high altitude emerged and spread. People from Tibet to this day have higher frequencies of genes, such as the oxygen-sensing EGLN and transcription factor EPAS1, which are associated with lower hemoglobin concentrations in the blood that life at height required.

Our bodies' adaptations to the environmental conditions we encountered during our migratory journeys help explain why we carry genes that heighten our risk of disease. Usually, genes that make it more likely we'll get sick die out over time, as people free of such genes out-reproduce people who carry them. But today a wide variety of genes raise our risk of illness, with nearly all diseases and health conditions having some genetic component.

Some of these genes likely persisted because they helped our ancestors survive the landscapes of the past. A mutation in the GDF5 gene, present in over 50 percent of Europeans and up to 90 percent in some Asian populations, for example, increases the risk of arthritis. When inserted into mice, it exhibits another effect as well: the mutation decreases the length of bones, leading researchers to conclude that the mutation is likely associated with short stature. Because short stature reduces heat loss, researchers hypothesize that the genetic mutation may have arose to protect our ancestors from the cold and frostbite they encountered on migrations into the north. Genes that promote inflammation and increase our risk of chronic inflammatory diseases such as heart disease and arthritis today may have emerged and spread because they helped us survive repeated exposure to food shortages and infections. A genetic variant that puts us at risk of a disease called phenylketonuria—subject of the common newborn screening test performed in hospitals—may also help us survive pathogenic fungi. It's more common in people who live in damp, fungus-rich Scotland. Gene variants that increase the risk of kidney disease may also help protect people from epidemics of sleeping sickness carried by the tsetse fly. Suggestively, people with higher rates of kidney disease are those with recent ancestry in Africa, where tsetse flies and sleeping sickness lurk.

Given the complex and circuitous path that lies between our genes and our behavior, it's unlikely that any one or even any group of genes could be definitively fingered as the source of our migratory impulse. Genes rarely provide instructions for a single trait, especially not for a complex behavior such as migration. And even when a single gene does code for a single trait, it doesn't express that trait in a straightforward fashion, but responds to switches and cues from the environment and from other genes. And yet at the same time, given our long history of migrations around the globe, it's unlikely that there's no genetic component behind the human propensity to move. So far one potential candidate has been found: DRD4 7R+. In a 1999 study, geneticists discovered that the frequency of the gene in different human populations correlated with how far from Africa they lived, with higher frequencies among those who'd moved the farthest. It's more common in nomadic peoples and is associated with openness to new experiences, attention-deficit disorder, and bursts of focused creativity.

Far from being fixed, our bodies are fluid. Their shapes, sizes, colors, and ability to withstand climatic variations are not locked in for generations by rigid blueprints. We shed forms and physiologies in favor of others, depending on the changing context in which we find ourselves. Our bodies have evolved, in other words, to "evade substantial 'genetic commitment' to local ecological conditions," write the anthropologists Jay T. Stock and J. C. K. Wells.

Such environmental mutability does not evolve in immobile creatures that live in static, unchanging environments. It evolves in creatures who migrate.

Our bodies are built for it.

~

Like butterflies and wolves, human migrants change the ecosystems they enter. People who decide to migrate are not a random cross-section of the population, like those found perusing the aisles of a grocery store or wandering around a train station. Whether it's money, skills, connections, or stamina, migration requires capital. Those with no capital, such as the very poor, cannot easily afford to undertake it. Nor can those whose capital derives from land ownership, aristocratic lineage, or titles. They have wealth and status, but they can't take it with them.

Instead, social scientists have found, migrants tend to be the kind of people who don't have big bank accounts or landholdings or titles but are rich in good health, skills, education, and social connections with people in other places. Their capital is portable. Demographically, in other words, they're the kind of people who are "the bedrock of successful communities," McLeman notes. They're working people from the middle classes, younger and better educated than their nonmigrant peers, and more likely to hail from societies on the middle rungs of economic development. They're healthier, too. Public health experts have documented what they call the "healthy migrant effect," that is, the fact that migrants experience lower mortality rates than the host populations they enter. It's especially striking considering that the newly arrived live in worse conditions and have less access to health care than the already resident, and most hail from poorer countries to boot. One study found that immigrants to the United States, Canada, the United Kingdom, and Australia had lower rates of chronic disease and obesity than comparable native-born residents. Most smoked less, too.

Their movements create social phenomena with their own self-perpetuating momentum. They move in stepwise fashion, from the countryside to the city, from cities across state borders, from nearby countries to ones farther afield, sometimes within a life-time, other times, like the monarch butterflies that take four generations to migrate from Canada to Mexico, over the course of generations.

The movements of pioneer migrants pave the way for others to follow. As they arrive, they strengthen migrants' social network, lowering the cost of migrating for others. Pioneer migrants like my parents, who knew not one soul in the United States before they arrived from Mumbai, helped bring over my cousins and aunts and uncles and even some of their friends, providing them with a place to crash, tips on finding jobs, and rides to the specialty grocery stores where they could buy oily jars of mango pickles and skinny green chilis. The remittances and other support they send tether them to the places they left like fragile silken threads.

Migrants bring new cultural practices, recipes, and ways of living and thinking into the societies they join, injecting novelty into insular populations, just as the migrant wolves of Isle Royale and the butterflies of Coyote Ridge did. And very quickly, unless the host society bars them, they integrate with the locals. Even as fire and fury are directed at newcomers and the economic and cultural disruptions they cause, our mongrel societies can and do rapidly assimilate the migrants among us. Within a generation, social, economic, and health indicators that distinguish immigrants from locals—the number of babies they have, the kinds of jobs they take, their levels of educational attainment, the diseases they suffer—converge. In one study of immigrants in the United States, all the immigrant-native differences that economists could discern vanished within a single generation.

~

The arrival of migrants in coming years will undoubtedly be disruptive to communities, even as they replenish them with new minds and bodies. In the past, human migrations have flowed more rapidly from east to west than from south to north, as evidenced in the pattern of genes in populations. In the new era, that circulatory pattern will likely switch, flowing from south to north along the gradient of our warming planet. The pace will be faster. Migrations will unfold over years and decades, not centuries and millennia.

But the next great migration will not unfold as an unstoppable physical phenomenon, like a cold front sweeping in from the north. There is no straightforward equation between environmental disruption and migratory effect. Those that occur suddenly and catastrophically, like floods and storms, might be expected to produce the biggest migratory effect, but they don't. On the contrary, studies of the relationship between migration and sudden flooding and storms have found only a weak correlation. Generally, in those situations people migrate only temporarily and not very far, often moving back to the places they left to rebuild after time passes.

One kind of environmental change that produces a detectable uptick in migration is drought. A study of three decades of data from 36 countries in sub-Saharan Africa, for example, found a correlation between rainfall shortages and increased rural-to-urban migration. Another found that a 10 percent increase in the number of communities experiencing drought correlated with a 10 percent increase in the number of people on the move. Some of the most prominent modern migrations have proceeded in the wake of drought. The Dust Bowl of the 1930s led to the migration of over 2 million people out of the plains states. Hundreds of thousands left their shanties to resettle in California, defying the California sheriffs who blockaded the border with Arizona to repel them. In the corridor of dry forests that wends along the Pacific coast of Central America from western Guatemala to northern Costa Rica, drought has unfolded in tandem with the growing ranks of people migrating from Guatemala, El Salvador, and Honduras, amassing on the United States' southern border as I write.

It is telling that slow-onset environmental disruptions leave a more detectable signal on migratory flows than do rapid-onset ones. Unlike storms and floods, which level their catastrophic effects all at once, droughts unfold gradually over time. First the rains become unreliable. Then they fail intermittently. Then comes a string of dry years. Like the goats on Mount Etna before the volcanic eruption and the elephants in Sri Lanka before the tsunami, the sons and daughters of farmers and fishers can detect the signs and understand that it's time to move on. Migration, in that case, is not a last-ditch escape from catastrophe or a zombie march over the ledge envisioned by some alarmists. It's a much more nuanced and adaptive response to subtle cues in the environment.

It's also mediated by politics. The mass exodus out of Syria, for example, was preceded by one of the worst droughts on record, during which crops collapsed and livestock herds perished. That led to high food prices, which forced rural people to flock to the cities. Between 2002 and 2010, Syria's urban populace grew by 50 percent, as its cities absorbed 1.5 million newcomers from the countryside. Many crowded into makeshift settlements, where political unrest against the corruption and neglect of the ruling regime built. An explosive civil war followed, which then led to waves of migration.

But the drought alone did not cause people to flee Syria. The failure of the political regime to stabilize food prices and provide food aid played as much of a role. So did the lack of adequate housing and jobs in the cities, and the regime's brutal response to the unrest that triggered the war. Elsewhere, drought may not influence migration at all. In the United States, increasing heat and aridity are unlikely to have much impact on migration patterns, because they have only small effects on agricultural profits, thanks to the resilience of the economic system.

When storms worsen, sea levels rise, and rains fail, picking up and leaving isn't the sole option. Societies may decide instead to ensure that people live in homes that can withstand the weather and that they can grow food under changing conditions. Settlement patterns, building codes, and agricultural practices can all be altered to enhance people's ability to stay. And the trigger for actions that improve resilience can be environmental disruption itself.

Water scarcity, which is commonly held up as a precursor to conflict and migration, has also led to hundreds of agreements between people across borders, including those whose governments are otherwise in conflict. Inhospitable weather, during the Little Ice Age, a period of several centuries during the Middle Ages, led Europe to abandon feudalism and ushered in the Enlightenment, the historian Philipp Blom writes. In response to desertification, the African Union has launched an effort to build a mosaic of drought-tolerant farms and forests on an eight-thousand-kilometer path across the continent, in an initative dubbed the "Great Green Wall."

~

For people like me who've lived their lives feeling out of place, the Heraclitan view of nature that has emerged in the past few decades provides a paradoxical sense of belonging. I remember the stab of envy I felt a few years ago, when I passed by two women at a local farmer's market. One, wearing a head scarf and a long gown, followed tentatively behind the other, a middle-aged woman in jeans and a sweater who confidently strode ahead, explaining the variety of tomatoes and lettuces available in the market. Given my city's program of resettling refugees, I assumed the first woman was a newcomer, and the second a volunteer helping her acculturate.

The idea of providing this sort of assistance, with its implicit recognition of one's insider cultural knowledge, immediately appealed to me. (I also get excited when anyone asks me for directions, something that happens only rarely.) But I quickly squelched the idea of volunteering. No foreign newcomer eager to acculturate would want a marginal American like me as his or her guide. Such volunteer work would join any number of other cultural activities I refrained from, like rooting for the home team and boasting about one's city or town.

But what I've come to understand is that in the broad view of human history, we're all migrants in every place we live, outside parts of Africa. Drawing the line between natives and outsiders based on some number of generations of continuous habitation is, in the end, arbitrary. Even Donald Trump is the child of an immigrant, like me. His mother migrated in 1930, sailing on a British passenger liner from her Gaelic-speaking home in Scotland's Outer Hebrides islands to New York City, where she took a job as a domestic worker. Her son and others like him shed their migrant histories like outgrown skins, skipping over the border between migrant and native without a look back. But their nativeness was just as provisional as my own.

I recognize that shifting migration from the margins of human experience to the center will not be comforting to many. Those of us who have been taught to expect stability feel entitled to an unchanging nature and our enduring place within it. But scientific findings have made it clear that migration is not an exception to the rule. We've been moving all along. And there's no singular factor that explains why, and that can be isolated and reversed to restore some mythical stasis.

Accepting that allowed me to see myself in a new way: as entitled to my patch of earth as anyone else. If anyone cared to ask, I would now call myself an American, with no extra adjectives to complicate it. And a Baltimorean, too, one with sufficient cultural knowledge to serve as a volunteer for any refugee newcomer.

~

The past seven thousand years of human migration unfolded under a globally stable climate, with the average global temperature ranging just 0.5 degrees Celsius. That has now changed. Since the industrial revolution, the average global temperature has risen by 0.8 degrees Celsius, bringing longer droughts, stronger storms, and more catastrophic wildfires. More of us than ever before will reach the threshold of the migration ratio, when our generation time eclipses the period of stability we can expect from the places we live in.

But as the next great migration dawns, the relevant question to ask is not why people migrate. Migration is a force of nature, rooted in human biology and history, along with that of the scores of other wild species with whom we share this changing planet. Over the long history of life on earth, its benefits have outweighed its costs.

The relevant question to ask is what we are going to do about it.

# 10

# THE WALL

The two migrants had set off from Iraq. Most likely relatives accompanied them, although I cannot be sure. Whether those relatives reached their destination, I also do not know. I do know that these two migrants ended their journey not far from where I stand, atop a hill on the Greek island of Lesbos, overlooking the shimmering blue Aegean Sea.

I'm in the corner of a local church cemetery, where the island's residents mark the graves of their dead in dense rows of monuments and etched marble slabs. Slinky cats lounge on the stone structures in various states of torpor. Some months earlier the migrants' bodies washed up on one of the island's beaches below. Locals carried them up the island's steep, bougainvillea-lined lanes to hand over to the gravedigger here, a wiry gray-haired man named Christos.

He'd received many such bodies by then. They had initially posed a dilemma for him, because the cemetery was reserved for the island's Christians, and while he did not know much about the migrants, he knew they were most likely Muslim. His solution had been to establish a kind of ghetto within the graveyard. He buried them in the cemetery's scruffy perimeter, where cemetery workers dump the graveyard's assorted debris into a rough pile and the grass grows wild.

The migrant graves were fresh when I visited, the spots where their bodies rested marked with a couple of shards of marble. No one knew the migrants' names, so the gravedigger had hand-painted what he assumed to be their ages on the shards with a bit of black paint. Some sympathetic locals who'd stopped by after paying respects to the statelier graves of their relatives and friends had left behind a bouquet of plastic pink flowers and two stuffed animals of the kind you might find in a pharmacy's gift aisle. The toys rested on a marble shard, its smudged painted inscription half-submerged in the mud.

The boy had been about five. His companion, about seven. The two children had made it over one thousand miles, across violent borders, and through countries ravaged by war. But barred from official ports of entry, they'd been killed by less than twenty miles of Mediterranean sea.

~

Another migrant left from somewhere south of the United States, donning a blue-and-white American Eagle polo shirt for the journey north. He made it across the Rio Grande, the wide and shallow river that meanders along the border between Texas and Mexico. It was the desert beyond that felled him. Unlike the mesquite and cacti, his body could not manage its dehydrating heat.

The remains of migrants who die in the desert are easy to lose. Over 90 percent of the borderlands in Texas consist of private ranches, some of which are so large that vast swaths go unvisited for years. The biggest one is the size of the state of Rhode Island. And the desert rapidly metabolizes human flesh. Within a day, a dead migrant's body in the desert is visually unrecognizable.

Coyotes strip away the flesh; vultures peck out the eyeballs. The bones that remain, bleached white by the desert sun, are incorporated into prickly stands of cacti or dragged into rats' nests, where the calcium hones the rodents' incisors.

Against the odds, someone found the young man's polo-shirted body before the desert digested it. A local official ferried the body to the morgue. State law requires that unidentified bodies such as his be sampled for DNA, and the FBI called to help with identification. But the remote South Texas county where he'd ended up, the fifth poorest in the United States, had little budget or political will for all that. Migrants die in South Texas regularly. The U.S. Border Patrol has counted over fifteen hundred migrant deaths in the Rio Grande valley since 1998. There are too many for local authorities to manage. At the morgue, a pathologist removed the young man's clothes and sliced open his chest to check for signs of foul play. Then the pathologist sewed his body back up, stuffed his clothes into a small biohazard bag, and zipped both into a black body bag.

The official who runs the morgue drove the zippered-up black bag down some dusty dirt roads on the edge of town to his family's sprawling cotton farm. He drove past the ramshackle ranch house, draped with Christmas lights, where a distant cousin sat drinking his daily quart of whiskey. He drove past the rickety fence that encircled the ranch house next door, abandoned by its owners to smugglers who used it as temporary housing for the migrants and drugs they pirated across the border. Finally he pulled up next to a lone tree casting a dainty shadow on a clearing.

His family had been burying their dead here for generations, surrounded by acres of cotton fields, marking their graves with engraved granite arranged in rows. For years, he'd been digging holes in between the marked graves and dumping the nameless dead who passed through his morgue there too: homeless people, unidentified hospital patients still wearing their flimsy hospital gowns and tethered to medical tubing, migrants who'd died in the desert or drowned in the river. Some he'd put in Styrofoam containers, some in leftover caskets from the local funeral home. Others, like the blue-and-white-polo-shirted migrant, got stashed in their black body bags. Sometimes he'd mark the grave with a paper tag reading "Jane Doe" or "John Doe," which the landscapers who mowed the grass would then knock over. He had no other option. There's no public cemetery, and customers at private cemeteries didn't like their loved ones being buried near unidentified bodies.

Ten years passed between then and the day I saw the polo-shirted migrant. The sky was cloudless with a strong, steady breeze. A forensic anthropologist had raised funds to bring a team of student volunteers to excavate the unidentified graves at the cemetery in the cotton fields and identify the remains. They carried pickaxes and shovels and looked for shallow depressions in the ground that might indicate a body surreptitiously hidden underneath. The morgue official had kept no records or map. When the anthropologists found one, they called over the backhoe and driver they'd hired, who carefully dug into the earth, until the tip of the shovel scraped a coffin or a bag. Over the course of a week, the team had exhumed nearly two dozen unmarked bodies.

As they lifted the black plastic bag encasing the young man's body out of its grave, I could see the shape of his rib cage. The plastic had gone brittle and clung to his body, which was flattened and dehydrated. They carried the bag to the prickly dry grass and laid it gently down in the shade of the tree. There a few donned masks and gloves to slice open the bag with a knife. It would take months if not years for them to identify the remains. They'd have to extract DNA from the bones, analyze it, upload the sequence to a public database, and then wait for the family of the missing to find it.

The first thing they'd do was try to identify the gender and age of the body. If they could do that, they could get a decent sense of whether the body was that of an abandoned hospital patient or a forgotten homeless person or a migrant lost in the desert or drowned in the river. But even that could take hours. Many of the bodies were lifted out of their unmarked graves in half-liquid, half-solid form.

A knot of us stood upwind, avoiding the stench. The backhoe hummed steadily in the background, probing the ground behind us.

The lead anthropologist gingerly examined the contents of the bag. Scraps of grungy fabric and decomposing tissue clung to the skeleton, cushioned in a bed of dark black soil. He pulled out a plastic biohazard bag nestled in the crook of the skeleton's neck. It must be the corpse's interior organs, he muttered, extracted during his autopsy.

But the plastic biohazard bag nestled in the crook of the young man's neck did not contain the remains of his kidneys or his heart. Instead, it contained the dead man's clothes: the polo shirt and a single white tube sock. The clothes were enough to determine who he was, at least in outline: a young man who'd hoped to start a new life across the border. He'd made it as far as a cotton field in South Texas, thirty miles north of the U.S.-Mexico border.

∾

On about 3.6 percent of the planet's surface, geographic barriers prevent wild species from migrating as effectively as the desert borderlands barred the polo-shirted young man. Take, for example, the mosaic-tailed rat, which lived on tiny Bramble Cay, an uninhabited island on the northern edge of the Great Barrier Reef off the coast of Australia. Increasingly violent storm surges steadily wiped out the island's plant life. But like other terrestrial creatures living on remote islands, or at the top of mountains, the mosaic-tailed rat had nowhere to go. The rodents' numbers diminished. By 2002 there were only ten mosaic-tailed rats left on the island. A fisherman spotted one in 2009, but when scientists returned in 2016 to survey the island, they couldn't find even one.

In 2019, with 97 percent of the vegetation on the island destroyed, officials in Australia declared *Melomys rubicola*, the Bramble Cay mosaic-tailed rat, officially extinct. It was the first mammal we know of to be wiped out by climate change. Experts agreed it would not be the last.

The more potent barrier to wild species' movement is us. So far, our cities, towns, farms, and sprawling industrial infrastructure have swallowed up over half the planet's land surface. We transformed another 22 percent of the earth's habitable land in just the last decades, mostly by cutting down forests and turning them into farms, as a recent analysis of satellite images from 1992 to 2015 showed. Our massive footprint makes life impossible for so many wild species that an estimated 150 go extinct every day, speeding up the background rate of extinction by a factor of one thousand.

Species that have not lost their habitats entirely must move through a landscape disfigured by human developments. Black bears in the hardwood swamps of Louisiana must cross a highway to reach others in their population. Instead of striking out across the highway to find new mates, they've started to mate with those in their own cut-off group, becoming increasingly inbred. Cougars living in the mountains around Los Angeles must cross two freeways, including one with eight lanes of speeding traffic, to meet others of their kind. None of the cougars that scientists fitted with GPS collars could do it. Four died attempting the crossing, five turned back, and one was shot by police. Birds on the wing smash into industrial structures, each building regularly racking up corpses, like the half-dozen or so birds felled every week by the Thurgood Marshall Federal Judiciary building in Washington,

D.C. Migrating butterflies, lured off course by electric lighting, perish and flutter to the ground.

A 2018 paper in *Science* analyzed the movements of fifty-seven different mammal species outfitted with GPS devices over land-scapes rated according to a "human footprint index," which incor-porated data on human population density, the extent of built land, roads, nighttime lighting, and the like. New York City garners a score of 50, while the vast and wild tropical wetlands of the Brazilian Pantanal scores 0. The bigger the human footprint, the researchers found, the more constrained animal movements became. In places with the largest footprints, animals managed to cover as little as a third of the distance of those in places with little or no human impact.

∾

Besides the inadvertent obstacles of geography and industrial development, the next great migration must overcome purposeful barriers.

Before 2001 fewer than twenty of the invisible boundaries that define nearly two hundred nation-states were physically marked by fences or walls. Animals, winds, currents, and waves could freely travel across their imaginary lines.

In 2015 an unprecedented surge in construction of new border walls began. By 2019 newly built walls, fences, and gates had risen over sixty international boundaries, blockading the movements of over 4 billion people around the world. More borders are fortified by walls and fences today than at any time in history.

Tunisia has built a wall of sandbanks and water-filled trenches along its border with Libya. India and Myanmar have fenced their borders with Bangladesh. Israel has enclosed itself with razor wire, touch sensors, infrared cameras, and motion detectors. Hungary's fence along its border with Croatia, built by prisoners, delivers electric shocks to any migrant foolhardy enough to touch it. Security officials patrol the barrier, tear gas canisters in hand.

Austria has built a fence along its border with Slovenia. Britain plans another one along the channel separating it from France. Norway has fortified its border with Russia. In the United States, the hundreds of miles of sixteen-foot-high concrete and steel walls marking the southern border would be extended with even taller, longer, more impregnable walls, U.S. president Trump insisted, perhaps even the entire length of the two-thousand-mile border.

Walls don't necessarily function as the impregnable barricades they're meant to. In one study, for example, researchers set up camera traps along the U.S.-Mexico border, tracking the movement of people and wild species across open stretches and comparing their movement across stretches blocked by border walls. The walls effectively deterred the pumas and coatis. According to conservation biologists, the extended walls proposed for the U.S.-Mexico border, for which the government has waived scores of environmental regulations, will endanger the life-saving movements of most of the ninety-three species that live on either side of it. But in the study comparing open and walled stretches of the border, the walls had no effect on people's movements. Whether crossing the border means scaling a wall or not, people keep moving, regardless.

If border obstructions fail to stifle movements, they do effectively deflect them. People on the move take more circuitous routes than they otherwise would in order to circumvent barriers, moving like water around a boulder in a stream. Attempting to bar migration, a European border official said, "is very much like squeezing a balloon. When one route closes, the flows increase on another."

But not all migrant routes are the same. People on the move choose the safest and most direct routes first. As those close off, people get diverted into more exposed territory. They are more likely to walk deeper into the desert, launch their boats into rougher waters, and climb higher into the mountains. They are more likely to hire smugglers. Migration continues, but in a deadlier form.

Between 2015 and 2018, European officials erected a wide array of barriers to prevent people from migrating across the Mediterranean Sea to seek refuge in Europe. The number of people who crossed fell from over 1 million to less than 150,000. But the deadliness of the migrant route skyrocketed. Migrants ventured out into rougher waters under the thumb of more brutal smugglers. Fewer rescue operations were available to aid them. In 2015 one migrant died on the sea route to Europe for every 269 who arrived. In 2018 one migrant died for every fifty-one who arrived.

Crackdowns on land routes into Europe have had a similar effect. In 2016 European Union officials targeted migrants from West Africa who crossed from northern Niger into Libya to reach Europe. Smugglers were arrested; their vehicles confiscated; over two thousand migrants who reached the Niger-Libya border were deported. The flow of migrants traveling across Niger into Libya dropped precipitously. European Union officials gloated over the stellar results.

But the flow of people had simply shifted, westward out to sea and eastward into the desert. Every month an estimated six thousand people continued to head from West Africa toward Europe by land, but they crossed instead from Niger into Chad. The new route exposed them to more remote areas of the Sahara Desert. Their vehicles broke down in the 110-degree-Fahrenheit heat. Fearful of encountering police officers or soldiers, smugglers abandoned migrants to die of thirst. In the first eight months of 2017, smugglers ditched more than a thousand migrants in the Sahara. And that's just the number of stranded migrants found alive. Aid workers presumed that the number of abandoned migrants who died of thirst in the desert "likely exceeds the number of those rescued."

Migrants from West Africa headed west, farther out to sea, as well. Migrants arriving on the black-and white-sand beaches of Spain's Canary Islands quadrupled between 2017 and 2018. To make the journey, some had undertaken the thousand-mile open-ocean crossing in wooden boats loaded with dozens of passengers.

Overall, between 1993 and 2017, over 33,000 people died trying to migrate into Europe. Between 1998 and 2018, as many as 22,000 may have died trying to cross the U.S.-Mexico border into the United States.

The true figures are probably much higher. "I would say for every one we find, we're probably missing five," one law enforcement official in South Texas said.

~

Ghulam Haqyar, who had left Herat province in Afghanistan with his wife and four children after one of Haqyar's colleagues was murdered, finally did reach Europe. He was even able to salvage one of the German-language textbooks the family had carefully carried with them over the mountains and the sea, in preparation for the new lives they'd start in Germany.

Ahead of them, to the north and west, lay 1.6 million square miles of the European continent, comprising over two dozen countries that had maintained open borders since 1985. Streams of migrants like Haqyar who landed in the southern border countries such as Greece or Italy continued their journeys north unbothered by border authorities at checkpoints demanding papers, heading into the more prosperous parts of the continent, where they could apply for asylum and find jobs, housing, and social connections.

But by the time Haqyar's family made it over the Mediterranean, the borders had closed. Facing hundreds of thousands of newcomers, the governments of Europe changed their minds about their open-borders agreement. By 2016, officials had erected border checkpoints around Austria, Denmark, France, Germany, Norway, and Sweden. The European Union paid its bordering countries—Libya, Tunisia, Morocco, Turkey, and Egypt—to intercept and turn back migrants before they could reach Europe.

The border closures trapped tens of thousands of migrants in Europe's southern countries. Thousands camped out at Greek ports and in the mud and rain along Greece's northern boundary, hoping the border might crack open enough to let them pass. As the weeks dragged into months, reporters captured images of desperate migrants threatening to throw their own babies into the sea. Others hung themselves. Finally, after weeks of increasingly disturbing headlines, the Greek military razed migrants' ad hoc encampments on the borders and at the ports. Soldiers rounded up the migrants on buses and deposited them in hastily built military-run camps, where they'd be out of the public eye while European policy makers figured out what to do with them.

Haqyar and his family, along with eight hundred others, ended up in a military-run camp built on an old gravel parking lot in one of the hottest, driest parts of the country, about a three-hour drive from Athens. The soldiers provided them with a canvas tent, one visited nightly by snakes and scorpions, then made themselves scarce. They spent most of their days there ensconced inside their own tents, air-conditioned unlike the others. Outside, the soldiers' confused and traumatized charges wilted in the sun. No one told Haqyar or anyone else trapped in the camp how to apply for asylum. No one told them how long they'd be held there. Volunteer doctors watched as suicide rates and episodes of acute psychiatric illness spiked.

"I can't name one person here who isn't losing their mind," muttered one of Haqyar's fellow camp residents, a journalist who had fled Kabul with his wife and four children. "Even my small girl says, 'Daddy, Afghanistan is better than here,'" Haqyar added.

Some migrants have escaped the military-run camps. In Athens, a group of activists turned an abandoned schoolhouse into an ad hoc group home, where families from Syria and Afghanistan stuck in Greece could sleep on classroom floors, using blankets wedged between old student desks as walls. Conditions there weren't much better than in the camps. A local psychiatrist volunteered to visit twice a week to dispense medical advice from a first-floor classroom at the end of the hall equipped with a few dented desks and chairs. His well-meaning medical advice was no match for the array of ailments the migrants suffered. The evening I stopped by, a continuous stream of men, women, and children wandered in and out, complaining of heart palpitations, asthma, and strange and worrying rashes, their faces spotted with angry red lesions from an ongoing chickenpox outbreak. The psychiatrist had little besides a small supply of donated medicines to offer. He was easily rattled, at one point bellowing so angrily at one of the other volunteers that she rushed out of the room in tears.

At the military-run camp, Haqyar set up a makeshift school in one of the tents, where the dozens of children living in the camp could at least go through the motions of education. Their parents, exhausted from staying up all night beating snakes away from their sleeping babies, rested fitfully in their stifling tents in the shadeless sun, battling paralyzing feelings of abandonment and neglect, the scent of leaky Porta Potties wafting through the still air. The children, meanwhile, took turns paging through their sole book, the German textbook that Haqyar had ferried hundreds of miles across mountains and seas in hopes of a different future.

～

It is possible to argue that Greece had relatively few resources to provide for the humanitarian needs of migrants inadvertently stranded within its borders. Greece had been mired in a crippling economic crisis since 2008. Public hospitals lacked adequate supplies even for longtime residents, let alone a sudden stream of newcomers. Conditions in the country's immigration detention camps were so dire that the European Court of Human Rights ruled in 2011 that they amounted to torture.

Deprivation is also, for some political leaders, a matter of policy.

Their crude logic is that their societies' generous public services act as attractants. There's plenty of evidence that that isn't true. If it were, people from poor countries would steadily empty into rich countries to which they have access. They don't. People from Niger, for example, can freely move to Nigeria, which is six times richer. People from Romania can freely move to Sweden, which is six times richer, too. Neither Niger nor Romania has depopulated as a result. In fact, most of the world's migrants move from one developing country to another, that is, between countries where the range of available public services varies little.

Still, on the theory that withholding society's riches will deter people from migrating, many countries in Europe exclude people without official documents from services that are freely available to locals. In six European Union countries, undocumented migrants are entitled only to emergency health care. In twelve other countries, they are excluded from primary and secondary care. Migrant children without official documents are not offered even rudimentary protective measures, such as vaccinations.

As a results the health of migrants, which starts off superior to those in the host societies they enter, steadily erodes. One survey of people who'd fled Iraq and ended up in the Netherlands found that the rate of psychiatric disorders and chronic physical ailments grew in direct relation to the length of time they'd been in the Netherlands, waiting for papers.

By 2019 in the United States, the Trump administration's deterrence policies went beyond deprivation to the purposeful infliction

# Migrant deaths

■ European Union and United Kingdom

■ Countries that have agreed to detain migrants in camps

□ Countries that have received funds from the EU to prevent migrants from reaching Europe

More borders are fortified by walls and fences today than at any time in history. While barriers may fail to deter migration, they effectively deflect it, often into more deadly territory. Between 1998 and 2018, as many as 22,000 people may have died trying to cross the U.S.-Mexico border into the United States; between 1992 and 2019, over 36,000 people died trying to migrate into Europe, as this map depicts. Circles are proportional to the number of fatalities. The dots represent detention centers and camps; the lines enclose the area that European officials police to repel migrants.

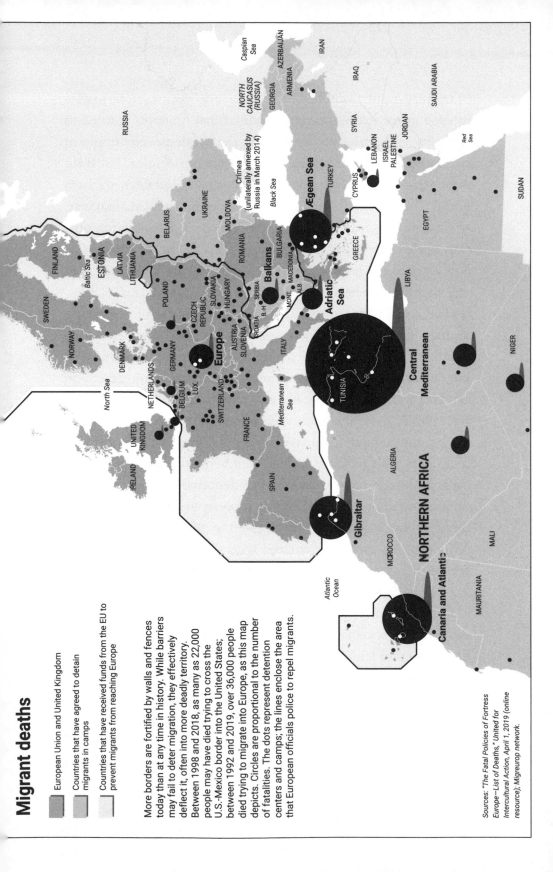

Sources: "The Fatal Policies of Fortress Europe—List of Deaths," United for Intercultural Action, April 1, 2019 (online resource); Migreurop network.

of trauma. A policy dubbed "zero tolerance," for example, implemented in the spring of 2018, required that migrants entitled to apply for asylum be prosecuted for the misdemeanor crime of crossing the border irregularly first. These prosecutions, in turn, meant that the children whom migrants had brought with them in their flight from poverty and violence were detained separately, while the adults made their way through court. Immigration officials herded over 2,300 children into camps encircled with chain-link fencing, where even the breastfeeding infants and toddlers in diapers among them would have to fend for themselves.

The United States is not alone in jailing children for violating immigration rules. More than one hundred other countries do it, too. But few have ever implemented the separation of families and the detention of children so carelessly. In immigrant detention centers, authorities physically tore children from their parents' arms. Women waiting to file their claims were tricked into briefly leaving their children to have their photographs taken, only to find upon their return that their children had vanished. If they made it through court without being deported—a feat, given that they rarely had legal representation or even anyone speaking their language during court proceedings—the government could not guarantee that they'd be reunited with their children. Some had been sent hundreds of miles away. Some were deported. Others were farmed out to relatives. Government officials admitted, in leaked emails, that they did not keep track of the children's whereabouts in any systematic fashion. "No, we do not have any linkages from parents to [children]," one official wrote to another. "We have a list of parent alien numbers but no way to link them to children."

Critics, noting the unsanitary and overcrowded conditions in the detention centers and the unwashed, sickened, and traumatized children within them, complained that the policy amounted to state-sponsored kidnapping and child abuse. But President

Trump claimed that separating parents from their children would deter migrants. "If they feel there will be separation," he explained, "they don't come." The government's own data suggest otherwise. Before it was implemented across the U.S.-Mexico border, the "zero tolerance" policy had been rolled out along the border near El Paso, Texas. Between July 2017 when it began and November 2017 when it ended, the number of families caught trying to cross the border did not decline. On the contrary, it rose by 64 percent.

~

Jean-Pierre's death-defying journey through more than a half-dozen countries and the wilds of the Darién Gap could have ended in Orlando, Florida. Thanks to a backlog of hundreds of thousands of cases in immigration courts, and the government's refusal to hire the judges and others needed to handle them, the asylum hearing he'd been promised would probably not be held for years. But before he and his family could start to settle, a low rumble of portents suggested that it was time to leave once more.

The Trump administration enacted policies to stymie migrants' right to claim asylum. Along the country's southern border, a policy known as "expedited removal" empowered border officials to decide for themselves whether people seeking refuge deserved to have their cases heard by a judge, allowing them to summarily deport those deemed fraudulent or deceptive. Under another policy known as "metering," border officials arbitrarily restricted the number of asylum applications they'd accept, forcing migrants to wait for weeks to even submit an application. A policy known as the "migration protection protocols" required that migrants await their asylum hearings, sometimes for years, in Mexico rather than in the United States. Bilateral agreements, forged under the threat that the United States would cancel hundreds of millions of dollars in foreign aid, would allow the United States to reject asylum claims from anyone who had traveled through El Salvador, Guatemala, or Honduras, and hadn't applied there first.

Other policies targeted immigrants already settled in the country. Migrants who'd been living in the United States for far longer than Jean-Pierre started disappearing. In Ohio, immigration officials scooped up a businessman and deported him to Jordan. He had been living in the United States for nearly forty years and had raised four daughters. He left the country with nothing more than the clothes on his back and a few hundred dollars in his pocket. In Connecticut, they picked up a couple and deported them to China. They'd lived in the United States for nearly two decades and had been running a local nail salon. They had to leave their five-year-old and fifteen-year-old sons behind. In Iowa, a teenager who'd lived there since the age of three was deported to Mexico. He was murdered shortly after arriving.

While previous administrations had captured and deported migrants living in the interior of the country before, they'd primarily targeted those who'd been convicted of crimes. In a single year, the number of migrants living in the interior who'd been arrested shot up by 40 percent. The majority had no criminal convictions at all. Their sole violation consisted of a lack of valid immigration documents.

Even legal immigrants and those who had become citizens fell prey. Under a new "public charge" rule, the Trump administration announced it would penalize legal immigrants for using public services such as food stamps and housing assistance by denying their applications for permanent legal status. Citizens would be subject to denaturalization if their papers were found to be faulty.

People from Haiti were subject to particular scrutiny. Officials in the White House watched approvingly as the Dominican Republic ousted hundreds of thousands of Haitians. In 2013 a government tribunal in the Dominican Republic had ruled that anyone who could not prove that at the time of their birth their parents were citizens would henceforth be considered foreigners and be subject to deportation. With one stroke, they'd abruptly rescinded *jus soli*, or birthright citizenship—the right to citizenship in the country of one's birth—from hundreds of thousands of people, most of whom hailed from the country's next-door neighbor, Haiti. One of President Trump's top immigration officials praised the "clarity" of the new policy. The Trump administration hopes to end birthright citizenship, too. Trump promised as much on the campaign trail.

The Dominican Republic started expelling Haitians en masse in 2015. By 2018 it had kicked out eighty thousand, many of whom ended up living in filthy makeshift camps along the border.

Even before the Trump administration rescinded Haitians' immigration status, Haitian neighborhoods emptied. Haitian churches were abandoned. Two hundred people used to attend the Haitian Methodist Ministry service in San Diego, but by the summer of 2017, only thirty or so remained. Community workers who helped drive Haitians to doctors' appointments arrived at their houses to find them abandoned. Later they'd get a text from Canada.

The migrants streamed north. Between the spring of 2017 and the spring of 2018, more than twenty thousand fled the United States to seek refuge in Canada. They didn't get far. At the official border crossings, Canadian authorities often handed them over to the U.S. immigration officials they'd fled. If they hadn't first applied for asylum in the United States—regardless of how patently fruitless that endeavor had become—Canada would not hear their claims. Once back in the custody of U.S. immigration authorities, the fleeing migrants were split up, imprisoned, and deported. U.S. officials sent the father of one Haitian family who'd attempted to claim asylum in Canada to the local county jail. They sent his pregnant wife and their small children to a run-down hotel used for people newly released from prison. "They didn't have transportation, they don't have money, and they are paying to stay at this hotel, not knowing what to do," a local woman who tried to help the distraught family noted. The marooned kids had only two pairs of thin socks each to make do in chilly upstate New York.

To avoid their fate, other fearful asylum seekers set off into the snow-covered woods that blanketed many of the unguarded portions of the border. One man who'd been turned away at the official border crossing into Canada spent nine hours wandering in the woods between the United States and Canada, during which the temperature plunged to negative 15 degrees. Police found him, barely conscious, the next morning. He had fallen through a thin layer of ice into a freezing-cold river. His feet were swollen and covered in blisters. They took him to the hospital, where they handcuffed him to the bed, and then, when he recovered, whisked him off to a detention center. He still dreams about the forest, he says. In the dream, not unlike in his waking life, "I'm screaming and no one is around for my rescue," he said.

Jean-Pierre and his family abandoned their attempt to secure asylum in the United States and boarded a bus to Plattsburgh, New York. From there, they caught a cab to Roxham Road, a sleepy residential lane in Champlain, New York. You wouldn't know it from looking at it, but the Canadian border bisects this road. About a mile down, past the clumps of spotted horses listlessly eating hay and a couple of run-down farms, the road ends, as if at a dead end. Beyond lie a few boulders, a five-foot-wide ditch, and a small grassy clearing, which is all that lies between the United States and Canada. The country road it interrupts doesn't make a big deal about it. After the grassy clearing, it picks back up, the same as before, continuing nonchalantly through several more miles of farmland.

During the summer of 2017, taxicabs arrived in a steady stream, disgorging carloads of asylum seekers and their hastily packed luggage, turning the quiet country road into something more like a scene outside JFK Airport. Crossing the international border here is technically illegal. But unlike at official entry points, Canadian officals would adjudicate the claims of those who entered the country from Roxham Road. As the number of border crossers grew, Canadian border officials set up white tents in the grassy clearing to process the newcomers. Once they stepped over the border, they'd be beholden to the bureaucracy that would have to hear their claims, which could take weeks or months. They wouldn't be able to backtrack, not even the few steps back to the road. One family, in their haste to escape the United States, had left their luggage a few steps away, beside the taxicab that had brought them to Roxham Road. They had to leave it behind, entering their new life with nothing except the clothes on their backs.

Jean-Pierre and his family were among them. They spent twenty-four hours getting processed in a tent on Roxham Road, then Canadian officials bused them to a makeshift shelter at an old Olympic stadium in Montreal. They stayed there for two weeks. When I met him, he'd found a single-room flat in the basement of a dilapidated apartment building on the edge of Montreal, where his family of three awaited a hearing with a judge, who would decide their fate.

A pilly brown blanket had been tacked across the single window, high up on the wall, to obscure its thin light. The family's single bed occupied much of the dim space, leaving room for just a small table with a couple of folding chairs, from which Jean-Pierre growled answers to my questions.

His journey is still not over. The local volunteer who is helping him with his asylum case tells me later that the judge is unlikely to grant him permission to stay in Canada. Refugee status is reserved not just for those who are most deserving of refuge but for those who appreciate the country's magnanimity as well. Jean-Pierre, after all he's been through, will not perform the role of a grateful refugee. He is too angry and depressed.

~

As the myth of a sedentary past evaporates, a previously obscured question emerges: not why people migrate but why their movements inspire terror.

Xenophobia is not a uniform response to migration. It does not surge wherever unfamiliar populations collide, social science researchers have found. It is not more common in places in which there's a high proportion of newcomers. Nor does it emanate from the economically distressed, who may feel most threatened by the presence of newcomers. (The people who voted for Donald Trump, for example, whose most coherent position is his unapologetic opposition to foreigners, earn wages that are on average $16,000 above the median in the states in which they live.)

One study suggests that xenophobic outbursts are associated with a society's specific geopolitical history. Another proposes that it's the nature of settlement patterns that inflames fears of foreigners, specifically the relative size of differentiated populations and their level of segregation. Yet another speculates that xenophobic eruptions derive from the diminishment of restraints on hatred of foreigners, such as corporations' rising and falling demands for immigrant labor. When powerful actors need immigrant labor, xenophobia diminishes; when they don't, it flourishes.

One telling study analyzed the counties and states that voted for Donald Trump in 2016 by a measure called the "diversity index," that is, the probability that any two randomly chosen people differ by race or national origin. The researchers found that the antimigrant politician found his greatest support among people living in places experiencing a rapid influx of people who'd been born elsewhere. The states that Trump won were not especially diverse. The diversity indexes in those states were lower than the national average, ranking in the bottom twenty of the fifty states. But in the counties that Trump won, the low diversity index is changing rapidly, rising nearly twice as fast as the national average.

Why would relatively homogenous but newly diversifying counties be especially receptive to xenophobic rhetoric? One possible explanation is their awareness of the burden exacted by the novel influx of newcomers. The early days of any transition are usually the most challenging. And newcomers, especially if they are unexpected or arrive in large numbers, might overwhelm a community's absorptive capacity, pitting the interests of migrants against those of the locals. But in most places, such effects are likely to be temporary. Most communities can and do expand to accommodate newcomers. In plenty of others, there are enough vacant houses and unfilled jobs to absorb newcomers straight away. Between 2007 and 2017, 80 percent of all counties in the United States lost working-age adults.

Another possible explanation has to do with the optics of particular immigrant settlement patterns. An influx of newcomers is more conspicuous in a relatively homogenous place than in a more diverse one. It's even more noticeable if its pace is quicker than elsewhere.

Conspicuousness satisfies a fundamental condition for antimigrant sentiment. For xenophobia to flourish, natives must be distinguishable from migrants. In social psychology experiments, subjects made aware of a border between insiders and outsiders will rapidly bond with those on their own side and reject those outside it. They'll judge insiders to be fairer than outsiders. They'll describe insiders as having broadly positive traits, and outsiders as possessing broadly negative ones. They'll notice variations between insiders but not among outsiders.

The line between outsider and insider does not have to accurately correspond with shared interests or meaningful characteristics between people on one side and those on the other. Awareness of a border triggers biases regardless. In social psychology experiments, researchers have divided subjects into groups based on arbitrary grounds such as a coin toss, or the color of the T-shirts they are wearing, or their preferred ice cream flavor. It makes no difference. Subjects will exhibit bias toward those on their side and discriminate against those on the other side.

Unless policies and circumstances conspire to make it otherwise, the border between natives and migrants can be nebulous. Newcomers arrive and quietly melt into local populations, even as social panics about them rise and fall. Nativeness and migrantness are not permanent states of being: they pass over us like bands of light and shadow. All of us who live outside sub-Saharan Africa—and many of those who live there as well—share a migratory history on some timescale or another. In the United States, nearly a third of us are less than one generation removed from an act of international migration. Every year 14 percent of us move from one part of the country to another, crossing borders into states with different laws, different customs, and different dialects, some of them as distant from each other as New York City and Casablanca or Cartagena.

Only occasionally does the fact of our continuous movement rise into public consciousness, which may be why xenophobia erupts only sporadically. In the counties that Trump won, a peculiar settlement pattern happened to turn migrants into the visual equivalent of the bright purple flowers atop tall stands of purple loosestrife. It raised people's awareness of migrants in distinction to natives, elevating the line between insider and outsider. The spectacle of border walls and the brutality of deprivation policies against migrants have the same effect. The images of migrant children locked up in cages, or migrants camping in the mud along the border or crammed inside abandoned Olympic stadiums, paint a wide, bright line, marking the difference between the natives and the foreigners for all to see. Without such spectacles elevating the distinction between migrants and residents, migration happens underneath our notice, like the circulation of blood through our veins. The distinction between natives and migrants that might alert us to it fades to the point of invisibility.

~

If the xenophobic policies and practices barring the next great migration arise from demographic vagaries and spectacles that happen to inflame our biases against outsiders, yet more questions arise. Why are we so alert to group distinctions and so ready to shun outsiders?

According to one theory, this tendency may have evolved as an immune response. Outsiders may not steal our jobs or commit more crimes or even be readily distinguishable from us, but in the era before modern medicine, they did pose a potential biological risk: they carried novel pathogens.

History is littered with examples of what happens when people introduce pathogens to which they are accustomed into new populations that have never encountered them before. In the fifteenth century, Europeans started introducing the smallpox and measles viruses they'd been living with for centuries into Native American populations. Over the following decades, Native American populations nearly collapsed. Ancient Rome's malaria posed such a mortal threat to outsiders that the Romans coined a saying: "When unable to defend herself by means of the sword, Rome could defend herself by means of the fever."

Suggestively, ethnocentrist and xenophobic tendencies do seem to correlate with the presence of pathogens in our environment and our awareness of them. In places where pathogens abound, such as in the tropics, people have formed more ethnic groups than they have in cool and temperate places, where the pathogen load is lighter. People who feel more vulnerable to infectious diseases express more xenophobic and ethnocentrist attitudes than people who feel less so. In experimental studies, simply heightening subjects' awareness of pathogens, by providing information about new strains of influenza, activates the xenophobic impulse. After being so informed, subjects express more xenophobic and ethnocentric sentiments.

But if xenophobia evolved as a kind of immune defense, it's a crude one. Fever is an ancient, primitive, and nonspecific immune defense that we share with almost every other vertebrate and even some invertebrates. In some cases it helps reduce the replication of microbial intruders. The body detects the presence of a microbial outsider, and blood rushes to its core, kicking the immune system into action and creating a hostile, scorching hot atmosphere for the intruder. But at the same time the heat stress destroys the body's own tissues. Sometimes, what began as an immune defense turns into a self-destructive reaction, leading to seizures, delirium, and collapse. Xenophobic reactions are similarly primitive, nonspecific, and potentially self-destructive.

One way people express their xenophobic fears about other groups is by exaggerating their numbers and appetites. A 2018 study found that people in nineteen of twenty-eight countries in the European Union overestimated the proportion of immigrants in their countries by a factor of two or more. People in Bulgaria, Poland, and Romania, which have disproportionately fewer immigrants than other European countries, overestimated the number of immigrants in their countries by a factor of more than eight.

In another study, pollsters asked people how much government support immigrants receive compared to native residents. Almost 25 percent of people in France, nearly 20 percent of those in Sweden, and 14 percent of people in the United States estimated that immigrants receive twice as much government support as natives—which isn't true in any of those countries.

As with an out-of-control fever, these inflamed perceptions are unrelated to the nature of the supposed threat. They continue regardless of facts. Providing accurate information about the size of immigrant populations to those who overestimate them, a 2019 paper reported, "does little to affect attitudes toward immigration." The number of immigrants who arrive and local communities' capacity to absorb them play little role in the scale of people's negative response to immigrants, once it is triggered. "Just making people think about immigrants," one pollster commented, "generates a strongly negative reaction in terms of redistribution."

If the fever of xenophobia evolved as a kind of immune defense, perhaps it once helped protect us. It is no longer useful for that purpose. Modern medicine provides us with the insights and technology we need to protect ourselves from pathogens, whether we shun strangers or not. Still, the vestigial impulse to suspect outsiders lingers, lodged deep in our psyches. Politicians can harness its heat simply by pointing to a border between "us" and "them."

CODA

# SAFE PASSAGE

I met Sophia and Mariam a couple years ago in a cramped second-floor apartment in a run-down neighborhood in East Baltimore, where a local NGO had placed the two women together with their children. As a newly christened volunteer for the local refugee agency, I'd been handed a pile of folders about each refugee family in need of help. Instructed to pick one, I'd chosen them. We talked through a local translator, patched in through a cell phone. Mariam, who had fled Eritrea on foot, made it to a refugee camp just over the border in Ethiopia. Freed from the persecution of Eritrea's military regime, she spent most of her time hanging around, somewhat aimlessly. She is lithe, playful, and quick to smile. But living in a refugee camp had excluded her from the productive activities of society. She did not go to school. She did not have a job. Her main memory of her time in the camp, when I ask her, is of playing pickup games of soccer.

Sophia's track out of Eritrea curled toward the north. From Sudan she made her way to Cairo, where she scraped by along the margins. The small cross she wore dangling on a chain around her neck marked her as an outsider, excluding her from mainstream

Egyptian society. She took a job cleaning hotel rooms. But the heavy lifting damaged her back, and the botched surgery that followed left her incapacitated and unable to work. In yet another stroke of bad luck, doctors diagnosed her little boy, fathered by a fellow Eritrean on the run whom she'd met in Cairo, with a cancerous tumor in his left kidney.

But Mariam and Sophia had a path to a more secure future. Through the local offices of the United Nations' refugee agency, Eritreans in Cairo and in refugee camps could apply for refugee status. The agency would scan their faces and collect their fingerprints and biographical data. If the officers found them acceptable, they might refer their cases to some other country, which after conducting its own investigation into their backgrounds and biographies might find them suitably harmless and deserving. They might be allowed to move to a place where they could start making a home and a life for themselves. Every year, the agency resettles around 100,000 of the nearly 26 million refugees it recognizes.

Mariam and Sophia both applied.

They waited for nearly a decade before they were granted refugee status. The UN agency accepted their applications and referred their cases to officers of the U.S. Refugee Resettlement Program, which decided where they would, from then on, be allowed to live. Separately, they collected their belongings and boarded planes that would deposit them in their new homes.

They wanted to find jobs, they said. They wanted their children to be educated. Sophia's son, a tall, watchful boy, leaned on his mother's knee, his eyes wide and his expression serious. Mariam's daughter took an opposite tack, screwing her face into exaggerated expressions, touching my things, and climbing up onto my lap in a successful effort to charm.

As we sat together on the carpeted floor and pondered their prospects, Mariam brought out from their little galley kitchen plates of glistening strawberries, thinly sliced apples, and sliced oranges. The kids gathered hungrily around a platter of *injera*, the Eritrean sourdough flatbread, with steaming spiced lentils and curried potatoes mounded atop it.

Mariam and Sophia knew only a few words of broken English. They had no job skills to speak of. They were refugees in a society whose leaders called refugees "animals," "pests," and worse; and they were black women in a city so plagued by poverty and so ordered by race that living in one of its poor black neighborhoods curtails life expectancy by three decades. They had to care for two toddlers. They didn't know how to drive. Who would hire them? How would they manage to get to work if anyone did?

They had little family around to call on for support. The fathers of their children lived thousands of miles away. Mariam's partner had been resettled in Germany; Sophia's in Sweden. A framed photograph of a young woman was propped up on a small shelf. It was Sophia's daughter, who lived in Eritrea. She'd been a toddler when Sophia left. Now she was a teenager. Sophia hadn't seen her in years. Borders had cut through her family like a freeway through a forest, scattering broken pieces across the continents.

One recent evening in December, I picked them up to go see the Christmas lights in downtown Baltimore. After parking the car, we had to walk a few blocks in below-freezing weather, during which they described to me how in Eritrea they celebrated Christmas with a special meal at church and a round of visits to neighbors. Then the scene of electrified American excess that I'd brought them to see came into view. On this particular city block, locals had looped strings of twinkling lights from their windows, porches, and roofs, in between their row houses and across the narrow street to the row houses facing theirs. They'd crammed their small front yards with giant electrified candy canes, plastic snowmen waving their chubby arms, and piles of shiny gift-wrapped packages under sculptural Christmas trees built out of beer cans and old hubcaps. A woman dressed as Santa Claus handed out cookies to the crowd gathered to ogle the spectacle. At the end of the street, couples holding snowsuit-clad babies on their hips lined up to snap photos of themselves standing next to a man wearing a felted reindeer costume.

In the car, as we drove back to their apartment, the women were quiet. "Is nice," Sophia finally said, nodding. "American Christmas." I didn't know what to say. The candied red and white extravaganza challenged my own fledgling sense of cultural competency. I couldn't imagine that it made any sense to her—it hardly made any sense to me. I turned up the heat. Mariam's toes were numb because she had not worn any socks under her thin black sneakers.

We drove in silence until we reached their neighborhood a few miles away. Months would pass before they found work, Mariam taking a night job at an industrial Laundromat, Sophia cleaning a cafeteria. As I turned into the driveway, their building emerged out of the shadows.

Despite the strangeness of the night, the uncertainty of her future, the precariousness of the journey that had brought her to this unlikely destination, Sophia looked up at the sight of her building, as if it were unexpected, and whispered softly to herself, "My home."

∿

The fractured landscape that migrants move across can be repaired for both people and wild species.

Instead of expanding the borders of isolated parks and reserves, new conservation efforts are seeking to stitch together private lands, ranches, farms, and parks into wide, long corridors across which animals can safely move. The Yellowstone to Yukon Initiative, for example, has brought hundreds of conservation groups together to manage more than five hundred thousand square miles stretching southward from northern Canada, to ease wildlife movement across the entire expanse. A similarly ambitious project aims to protect millions of square miles of jaguar habitat across fourteen countries from Mexico to Argentina. Conservationists have pinpointed at least twenty places around the world, including biodiverse but highly fragmented locales such as the Eastern Arc mountains of Tanzania and the Atlantic forest of Brazil, where similar wildlife corridors could connect isolated fragments of protected lands into more than half a million acres of continuous forest across which species could freely move.

New infrastructure built for wildlife can ease their movement over the obstacles we've created. In Canada, grizzlies, wolverines, and elk march across wildlife bridges suspended above and below the Trans-Canada Highway. In the Netherlands, deer, wild boar, and badgers make it across railroad lines, business parks, and sports complexes thanks to six hundred corridors specially designed for them. In Montana, black bears, coyotes, bobcats, and mountain lions pad across more than forty wildlife crossing structures built over an interstate highway. Elsewhere conservationists have built tunnels for toads, bridges for squirrels, and ladders for fish. They've stitched together vegetation-dripping green roofs for birds and butterflies to rest on as they pass overhead. Together such efforts could create a kind of interstate network for wild species, creating seamless wildlife corridors over vast regions.

The ability to move is no panacea of course. Species that shift their ranges as their habitats vanish may end up exposed to more dangers rather than fewer. In Russia, Pacific walruses whose sea ice has melted now swim to distant rocky beaches to haul out. In the summer of 2017, wildlife filmmakers watched as the elephantine creatures climbed to the top of the rocky cliffs, plunging to their deaths on the beaches below, exhausted. Those that successfully shift their ranges may be condemned as "invasives." Wild species that have been criticized as unwanted intruders include the endangered freshwater turtle from Vietnam and China that successfully established itself in Hawaii; the Monterey pine trees, endangered in California and Mexico, that made it to Australia and New Zealand; the endangered barbary sheep that arrived on the Canary Islands; and the Sacramento perch, which spread through the western United States before going extinct in California.

Still, for the thousands of species now on the move toward the poles and into the higher latitudes, movement could be their best shot at surviving in the new era of climate chaos.

It is possible to envision a world in which people, too, safely move across the landscape. People seeking to move as the climate changes or as their livelihoods dry up don't have to risk being hunted down by Border Patrol agents or drowning in the sea or dying in the desert. International borders that now bristle with armed guards, razor wire, and border walls could be made softer and more permeable, more like the borders between, say, Massachusetts and New York, or between France and Germany. Initiatives such as the United Nations' Global Compact for Safe, Orderly and Regular Migration suggest a possible framework. The compact calls for countries to create more legal pathways for migrants in search of new livelihoods. It calls for countries to collect and share data on migrants and provide them with proof of their identity, so that migration can become more regular and orderly. It includes measures to make it easier for migrants to send funds and other support to the places they've left behind. And it calls for turning the detention of migrants into a measure of last resort instead of a reflexive first step.

The permeable borders that the compact imagines wouldn't absolve newcomers from the responsibility of obeying local laws and customs or erase the distinctiveness of local cultures. Rather, they would make migration safe, dignified, and humane. One hundred and sixty-three of the United Nations' 194 member nations have adopted the voluntary, nonbinding compact. In 2019 Portugal incorporated it into its own national immigration policy.

The militarized borders that bar human movement today are not sacrosanct. They're not fundamental to our cultures or histories. People in Europe started drawing borders around their countries only a few centuries ago. The British lawyer who established the borders around India and Pakistan marked them out over the course of just a few weeks. Even the highly contested border between the United States and Mexico was mostly permeable until just a few decades ago. Throughout much of our history, kingdoms and empires rose and fell with blurry edges, each culture and people shading gradually from one to the next. It's not that borders were open or closed. They didn't exist at all.

If we were to accept migration as integral to life on a dynamic planet with shifting and unevenly distributed resources, there are any number of ways we could proceed. The migration ratio will continue its inexorable approach, regardless. People like Sophia and Jean-Pierre and Ghulam will continue to move. We can continue to think of this as a catastrophe. Or we can reclaim our history of migration and our place in nature as migrants like the butterflies and the birds. We can turn migration from a crisis into its opposite: the solution.

~

We're driving along an unpaved and deeply rutted road on a piercingly bright day in the city of Tijuana, Mexico, looking for the wall.

Unlike other Tijuana neighborhoods with their jauntily painted exteriors and cheery window boxes, the neighborhood that abuts the border wall between Mexico and the United States has an ominous feel. The houses are shuttered. The neighborhood is notorious for being the site where drug lords dissolve in acid the bodies of those they've murdered. The wall itself exudes death. It's studded with hundreds of hand-painted crosses, left behind by locals to mark the lives of those who failed to overcome it.

As if in silent architectural protest to the wall's menacing effect, residents here have situated their homes' windowless backs to it, using the strip of cleared land between their homes and the wall as a garbage dump. It's a noxious river of old tires, empty Coke bottles, and discarded porcelain toilets, with the occasional pile of indiscriminate plastic factory waste. The scene is menacing, but no one is around except for the residents' barking guard dogs, so we leave the car running and approach the wall. My shoes sink an inch into gummy pale clay.

I clamber up a pile of old tires to peer over the wall to the other side. From this teetering standpoint, I can scan its length, wending east and west for miles, dipping into a valley and then disappearing over the crest of a far hill. I can see the tall slabs that have been erected in front of it, prototypes of the new border wall the U.S. president plans to build, lined up in a row facing south like some demented version of Stonehenge.

The wall dissolves into insignificance next to the mountain ranges all around. They extend for thousands of miles along the western coast of the North American continent, from southern Mexico to northern Alaska, forming a natural passageway for the bighorn sheep, the mountain lions, and the checkerspot butterflies, among other wild species, to move north and upward as the climate shifts. Regardless of the border and its barriers, of centuries of being condemned as invaders and feared as unnatural border crossers, migrants still come.

Somewhere in the distance, checkerspot butterflies emerge from their cocoons. Their delicate wings, spotted in orange, cream, and black, start to beat. The corrugated metal wall I'm peering over is only sixteen feet high. Checkerspots travel low to the ground, just six or eight feet above the desert plants and flowers they feed on.

When the moment comes, their slight bodies lift into the air.

# ACKNOWLEDGMENTS

The germ of the idea for this book lodged itself in my mind while I was sitting in a cramped office in Athens, Greece, where I interviewed Médecins Sans Frontières's director of medical operation support, Apostolos Veizis, about what I then called the migrant "crisis." He patiently but methodically exposed and shot down every assumption embedded in my neophyte questions. I owe a debt to him for that. The circuitous reconstruction of my ideas about migration and migrants that followed ultimately resulted in this book. The Pulitzer Center on Crisis Reporting made my reporting on asylum seekers in Greece and my visit with Veizis possible. Their support is invaluable to so many journalists. I'm proud to count myself among them.

I wrote the proposal for this book in a narrow, deep valley nestled in the Himalayas, listening to the sounds of a river roaring with glacial meltwater and gazing at the deodars slowly inching their way up the mountains. I couldn't have hoped for a more visceral and heart-stopping view of the drama of a world on the move and the urgency of addressing it. I thank Geetika Nigam and Ritesh Sharma for their gracious hospitality and the steady supply of egg parathas and chai they delivered to sustain me.

To write this book, I tapped the expertise of scholars in a wide variety of fields, from biogeography and conservation biology to genetics, anthropology, and the history of science. Camille Parmesan's work on species on the move was especially critical. She was generous with her time, explaining her findings, pointing me to the research of others, and helping me make the contacts that led to my first sightings of checkerspot butterflies. Spring Strahm,

Dave Faulkner, and Alison Anderson of the U.S. Fish and Wildlife Service shared their butterfly-chasing entomological expertise, as did Stu Weiss of the Creekside Center for Earth Observation. In Vermont, Jeff Parsons led me on a tour of the northern borderlands; in Boston, Pardis Sabeti and her colleagues explained their research deciphering the intricacies of human diversity; in Hawaii, Rebecca Ostertag and Susan Cordell shared their innovative research transcending the native-alien divide; in Texas, the forensic anthropologists Kate Spradley and Tim Gocha allowed me to tag along on their excavation of unmarked migrant graves. I thank them all.

Everywhere I went, I found migrants trapped in detention centers and in refugee camps, and those forced on the run who were willing to speak to me, despite the duress of living in the shadows. I am humbled by their journeys and grateful that they shared them with me. Heroic aid groups and activists, including Syrian-American Medical Society, Médecins du Monde, Frantz André in Montreal, and Pastor Dieufort J. "Keke" Fleurissaint in Boston, organized meetings and translations to make it possible.

I've cited only a portion of the scholars who were willing to talk to me. Mark A. Davis, Jonathan Marks, Warwick Anderson, Nils Christian Stenseth, Peder Anker, Hugh Dingle, Alan de Queiroz, and Martin Wikelski were especially generous and forthcoming. Several of them reviewed early versions of the manuscript and offered helpful suggestions, including Reece Jones, Betsy Hartmann, Matthew Chew, and others.

Anthony Arnove has supported my work in many ways for years. Michelle Markley has offered her deep insights. I'm proud to call them friends. Celia and Ian Bardwell-Jones hosted me in the beautiful house they built under Hawaiian volcanoes. Philippe Rivière of Visionscarto not only designed the maps in this book along with Philippe Rekacewicz, he offered critical editorial feedback, too. They gave more than I could have asked for. My agent, Charlotte Sheedy, and my editor, Nancy Miller, and her team at Bloomsbury supported me throughout the development of this book. It wouldn't exist without them. I thank them all.

When I first started writing this book, I expected to have to dig deep to unearth evidence of antimigrant science in today's politics. The 2016 election upended those expectations. As antimigrant rhetoric and policy surged to the forefront of our politics, the evidence I needed appeared in the news seemingly daily. Writing parts of this book became technically easier but psychically harder.

My growing circle of activist friends and allies helped me find light in the darkness. So did my own mongrel tribe of migrants and border crossers: Mark, who always listened, read it all, and took me sailing when I couldn't think about it anymore; Zakir and Kush, who are models of the kind of grace, engagement, and kindness that those inheriting the climate-disrupted world we've created will require; and my parents, who crossed the ocean for a new life and showed me what resilience and courage looked like.

# REFERENCES

Aaronson, Trevor. "Trump Administration Skews Terror Data to Justify Anti-Muslim Travel Ban." *Intercept*, January 16, 2018.

Anderson, Warwick. "Hybridity, Race, and Science: The Voyage of the *Zaca*, 1934–1935." *Isis* 103, no. 2 (2012): 229–53.

———. "Racial Hybridity, Physical Anthropology, and Human Biology in the Colonial Laboratories of the United States." *Current Anthropology* 53, no. S5 (April 2012): S95–S107.

Anker, Peder. *Imperial Ecology: Environmental Order in the British Empire, 1895–1945.* Cambridge, MA: Harvard University Press, 2009.

Bashford, Alison. *Global Population: History, Geopolitics, and Life on Earth.* New York: Columbia University Press, 2014.

Bendyshe, T. "The History of Anthropology: On the Anthropology of Linnaeus—1735–1776." In *Memoirs Read Before the Anthropological Society of London* (London: Trübner and Co., 1865).

Benton-Cohen, Katherine. *Inventing the Immigration Problem: The Dillingham Commission and Its Legacy.* Cambridge, MA: Harvard University Press, 2018.

Black, Edwin. *War Against the Weak: Eugenics and America's Campaign to Create a Master Race.* Washington, D.C.: Dialog Press, 2003.

Blunt, Wilfrid. *Linnaeus: The Compleat Naturalist.* London: Francis Lincoln, 2004.

Broberg, Gunnar. "Anthropomorpha." In Frank Spencer, ed., *History of Physical Anthropology.* London: Routledge, 1996.

———. "*Homo sapiens*: Linnaeus's Classification of Man." In Tore Frangsmyr, ed., *Linnaeus: The Man and His Work.* Berkeley: University of California Press, 1983.

Chamberlin, J. Edward, and Sander L. Gilman, eds. *Degeneration: The Dark Side of Progress.* New York: Columbia University Press, 1985.

Cheshire, James, and Oliver Uberti. *Where the Animals Go: Tracking Wildlife with Technology in 50 Maps and Graphics.* New York: W. W. Norton, 2016.

Chew, Matthew K. "Ending with Elton: Preludes to Invasion Biology." PhD diss., Arizona State University, December 2006.

Chitty, Dennis. *Do Lemmings Commit Suicide? Beautiful Hypotheses and Ugly Facts.* New York: Oxford University Press, 1996.

Crawford, Michael H., and Benjamin C. Campbell, eds. *Causes and Consequences of Human Migration: An Evolutionary Perspective.* New York: Cambridge University Press, 2012.

Crotch, W. Duppa. "Further Remarks on the Lemming." *Zoological Journal of the Linnean Society* 13, no. 67 (1877): 157–60

Crowcroft, Peter. *Elton's Ecologists: A History of the Bureau of Animal Population.* Chicago: University of Chicago Press, 1991.

Curran, Andrew S. *The Anatomy of Blackness: Science and Slavery in an Age of Enlightenment.* Baltimore: Johns Hopkins University Press, 2011.

D'Antonio, Michael. "Trump's Move to End DACA Has Roots in America's Long, Shameful History of Eugenics." *Los Angeles Times*, September 14, 2017.

Darwin, Charles. *The Descent of Man, and Selection in Relation to Sex.* 1871; reprinted New York: Penguin, 2004.

Davenport, Charles B. *Heredity in Relation to Eugenics.* New York: Henry Holt, 1911.

Davenport, Charles B., et al., eds. *Eugenics in Race and State*, vol. 2, *Scientific Papers of the Second International Congress of Eugenics, Held at the American Museum of Natural History, September 22–28, 1921.* Baltimore: Williams & Wilkins, 1923.

DeParle, Jason. "The Anti-Immigration Crusader." *New York Times*, April 17, 2011.

Desrochers, Pierre, and Christine Hoffbauer. "The Postwar Intellectual Roots of *The Population Bomb*: Fairfield Osborn's *Our Plundered Planet* and William Vogt's *Road to Survival* in Retrospect." *Electronic Journal of Sustainable Development* 1, no. 3 (2009).

Dingle, Hugh. *Migration: The Biology of Life on the Move.* New York: Oxford University Press, 1996.

Dobzhansky, Theodosius. "Possibility That *Homo sapiens* Evolved Independently 5 Times Is Vanishingly Small." *Current Anthropology*, October 1963.

Ehrlich, Paul. Interview by WOI-TV, April 24, 1970, YouTube, https://www.youtube.com/watch?v=yzwiraikxxg.

————. *The Population Bomb*. Cutchogue, NY: Buccaneer Books, 1968.

Ehrlich, Paul R., and John P. Holdren. "Impact of Population Growth." *Science*, March 26, 1971.

Elton, Charles S. *The Ecology of Invasions by Animals and Plants*. 1958; reprinted Chicago: University of Chicago Press, 2000.

————. "Periodic Fluctuations in the Numbers of Animals: Their Causes and Effects." *Journal of Experimental Biology* 2, no. 1 (1924): 119–63.

Fausto-Sterling, Anne. "Gender, Race, and Nation: The Comparative Anatomy of 'Hottentot' Women in Europe, 1815–1817." In Jennifer Terry and Jacqueline Urla, eds., *Deviant Bodies: Critical Perspectives on Difference in Science and Popular Culture*. Bloomington: Indiana University Press, 1995.

Finney, Ben. "Myth, Experiment, and the Reinvention of Polynesian Voyaging." *American Anthropologist* 93, no. 2 (1991): 383–404.

Frangsmyr, Tore, ed. *Linnaeus: The Man and His Work*. Berkeley: University of California Press, 1983.

Gelb, Steven A., Garland E. Allen, Andrew Futterman, and Barry Mehler. "Rewriting Mental Testing History: The View from *The American Psychologist*." *Sage Race Relations Abstracts* 11 (May 1986).

Gocha, Timothy, Katherine Spradley, and Ryan Strand. "Bodies in Limbo: Issues in Identification and Repatriation of Migrant Remains in South Texas." In Krista Latham and Alyson J. O'Daniel, eds., *Sociopolitics of Migrant Death and Repatriation: Perspectives from Forensic Science*. Cham, Switzerland: Springer, 2018.

Gould, Stephen Jay. *The Flamingo's Smile: Reflections in Natural History*. New York: W. W. Norton, 1987.

Gutiérrez, Elena R. *Fertile Matters: The Politics of Mexican-American Women's Reproduction*. Austin: University of Texas Press, 2008.

Harmon, Amy. "Why White Supremacists Are Chugging Milk (And Why Geneticists Are Alarmed)." *New York Times*, October 17, 2018.

Hartmann, Betsy. *The America Syndrome: Apocalypse, War, and Our Call to Greatness*. New York: Seven Stories Press, 2017.

Holton, Graham E. L. "Heyerdahl's Kon Tiki Theory and the Denial of the Indigenous Past." *Anthropological Forum* 14, no. 2 (2004).

Horowitz, Daniel. *The Anxieties of Affluence: Critiques of American Consumer Culture, 1939–1979.* Amherst: University of Massachusetts Press, 2004.

Jablonski, Nina G. *Living Color: The Biological and Social Meaning of Skin Color.* Berkeley: University of California Press, 2012.

Jones, Reece, ed. *Open Borders: In Defense of Free Movement.* Athens: University of Georgia Press, 2019.

Kessler, Rebecca. "The Most Extreme Migration on Earth?" *Science*, June 7, 2011.

Kirkbride, Hilary. "What Are the Public Health Benefits of Screening Migrants for Infectious Diseases?" European Congress of Clinical Microbiology and Infectious Diseases, Amsterdam, April 12, 2016.

Koerner, Lisbet. *Linnaeus: Nature and Nation.* Cambridge, MA: Harvard University Press, 1999.

Lalami, Laila. "Who Is to Blame for the Cologne Sex Attacks?" *Nation*, March 10, 2016.

Lam, Katherine. "Border Patrol Agent Appeared to Be Ambushed by Illegal Immigrants, Bashed with Rocks Before Death." Fox News, November 21, 2017.

Laughlin, H. Hamilton. *The Second International Exhibition of Eugenics Held September 22 to October 22, 1921, in Connection with the Second International Congress of Eugenics in the American Museum of Natural History, New York: . . .* Baltimore: Williams & Wilkins, 1923.

Lewis, David. *We, the Navigators: The Ancient Art of Landfinding in the Pacific.* Honolulu: University of Hawaii Press, 1994.

Lim, May, Richard Metzler, and Yaneer Bar-Yam. "Global Pattern Formation and Ethnic/Cultural Violence." *Science* 317, no. 5844 (2007): 1540–44.

Lindkvist, Hugo. "Swedish Police Featured in Fox News Segment: Filmmaker Is a Madman." *Dagens Nyheter*, February 26, 2017.

Lindström, Jan, et al. "From Arctic Lemmings to Adaptive Dynamics: Charles Elton's Legacy in Population Ecology." *Biological Reviews* 76, no. 1 (2001): 129–58.

Mann, Charles C. "The Book That Incited a Worldwide Fear of Overpopulation." *Smithsonian*, January 2018.

Marks, Jonathan. *Human Biodiversity: Genes, Race, and History.* New York: Aldine De Gruyter, 1995.

Marris, Emma. "Tree Hitched a Ride to Island." *Nature*, June 18, 2014.

Massin, Benoit. "From Virchow to Fischer: Physical Anthropology and 'Modern Race Theories' in Wilhelmine Germany." In George Stocking, ed., *Volksgeist As Method and Ethic: Essays on Boasian Ethnography and the German Anthropological Tradition*. Madison: University of Wisconsin Press, 1988.

Mavroudi, Elizabeth, and Caroline Nagel. *Global Migration: Patterns, Processes, and Politics*. London: Routledge, 2016.

McAllister, Edward, and Alessandra Prentice. "African Migrants Turn to Deadly Ocean Route as Options Narrow." Reuters, December 3, 2018.

McLeman, Robert A. *Climate and Human Migration: Past Experiences, Future Challenges*. New York: Cambridge University Press, 2013.

Montagu, Ashley. "What Is Remarkable About Varieties of Man Is Likenesses, Not Differences." *Current Anthropology*, October 1963.

Mooney, H. A., and E. E. Cleland. "The Evolutionary Impact of Invasive Species." *Proceedings of the National Academy of Sciences* 98, no. 10 (2001): 5446–51.

Moore, Robert, Lindsey Bever, and Nick Miroff. "A Border Patrol Agent Is Dead in Texas, but the Circumstances Remain Murky." *Washington Post*, November 20, 2017.

Mukherjee, Siddhartha. *The Gene: An Intimate History*. New York: Scribner, 2016.

Nathan, Debbie. "How the Border Patrol Faked Statistics Showing a 73 Percent Rise in Assaults Against Agents." *Intercept*, April 23, 2018.

Nicholls, Henry. "The Truth About Norwegian Lemmings." BBC Earth, November 21, 2014.

Normandin, Sebastian, and Sean A. Valles. "How a Network of Conservationists and Population Control Activists Created the Contemporary U.S. Anti-Immigration Movement." *Endeavour* 39, no. 2 (2015): 95–105.

Nowrasteh, Alex. "Deaths of Border Patrol Agents Don't Argue for a Longer Mexico Border Wall." *Newsweek*, November 28, 2017.

Osborn, Henry Fairfield. "Lo, the Poor Nordic!" (letter to the editor). *New York Times*, April 8, 1924.

Pierpont, Claudia Roth. "The Measure of America: How a Rebel Anthropologist Waged War on Racism." *New Yorker*, March 8, 2004.

Provine, William B. "Geneticists and the Biology of Race Crossing." *Science* 182, no. 4114 (1973): 790–96.

Queiroz, Alan de. *The Monkey's Voyage: How Improbable Journeys Shaped the History of Life*. New York: Basic Books, 2014.

Ramsden, Edmund. "Confronting the Stigma of Perfection: Genetic Demography, Diversity and the Quest for a Democratic Eugenics in the Post-War United States." London School of Economics, August 2006.

Ramsden, Edmund, and Jon Adams. "Escaping the Laboratory: The Rodent Experiments of John B. Calhoun and Their Cultural Influence." *Journal of Social History*, Spring 2009.

Ramsden, Edmund, and Duncan Wilson. "The Suicidal Animal: Science and the Nature of Self-Destruction." *Past and Present*, August 2014, 201–42.

Reed, Brian. "Fear and Loathing in Homer and Rockville, Act One: Fear." *This American Life*, July 21, 2017.

Reich, David. *Who We Are and How We Got Here: Ancient DNA and the New Science of the Human Past*. New York: Pantheon, 2018.

Ritz, John-David, and Aretha Bergdahl. "People in Sweden's Alleged 'No-Go Zones' Talk About What It's Like to Live There." *Vice*, November 2, 2016.

Rivas, Jorge. "DHS Ignored Its Own Staff's Findings Before Ending Humanitarian Program for Haitians." *Splinter*, April 17, 2018.

Roberts, Dorothy. *Fatal Invention: How Science, Politics, and Big Business Re-Create Race in the Twenty-First Century*. New York: New Press, 2012.

Roberts, Leslie. "How to Sample the World's Genetic Diversity." *Science*, August 28, 1992.

Robertson, Thomas. *The Malthusian Moment: Global Population Growth and the Birth of American Environmentalism*. New Brunswick, NJ: Rutgers University Press, 2012.

Rohe, John F. *Mary Lou and John Tanton: A Journey into American Conservation; Biography of Mary Lou and John Tanton*. Washington, D.C.: FAIR Horizon Press, 2002, https://www.johntanton.org/docs/book_tanton_biography_jr.pdf.

Schiebinger, Londa. *Nature's Body: Gender in the Making of Modern Science*. Boston: Beacon Press, 1993.

———. "Taxonomy for Human Beings." In Gill Kirkup, Linda Janes, Kathryn Woodward, and Fiona Hovenden, eds., *The Gendered Cyborg: A Reader*. London: Routledge, 2000.

Schmidt, Benjamin. *Inventing Exoticism: Geography, Globalism, and Europe's Early Modern World*. Philadelphia: University of Pennsylvania Press, 2015.

Shapiro, Harry Lionel. *The Pitcairn Islanders* (formerly *The Heritage of the Bounty*). New York: Simon and Schuster, 1968.

Sloan, Phillip. "The Gaze of Natural History." In Christopher Fox, Roy Porter, and Robert Wokler, eds., *Inventing Human Science: Eighteenth-Century Domains*. Berkeley: University of California Press, 1995.

Smethurst, P. *Travel Writing and the Natural World, 1768–1840*. London: Palgrave Macmillan, 2012.

Smith, Dylan. "Bannon: Killing of BP Agent Brian Terry Helped Elect Trump." *Tucson Sentinel*, November 18, 2017.

Social Contract. "A Tribute to Dr. John H. Tanton." YouTube, September 28, 2016, https://www.youtube.com/watch?v=cc2amo8oakq.

Spiro, Jonathan Peter. *Defending the Master Race: Conservation, Eugenics, and the Legacy of Madison Grant*. Burlington: University of Vermont Press, 2009.

Stenseth, Nils Christian, and Rolf Anker Ims, eds. *The Biology of Lemmings*. London: Academic Press for the Linnean Society of London, 1993.

Sussman, Robert Wald. *The Myth of Race: The Troubling Persistence of an Unscientific Idea*. Cambridge, MA: Harvard University Press, 2014.

Switek, Brian. "The Tragedy of Saartje Baartman." *Science Blogs*, February 27, 2009.

Tanton, John H. "International Migration as an Obstacle to Achieving World Stability." *Ecologist* 6 (1976): 221–27.

Taylor, Adam. "Who Is Nils Bildt? Swedish 'National Security Adviser' Interviewed by Fox News Is a Mystery to Swedes." *Washington Post*, February 25, 2017.

Thompson, Ken. *Where Do Camels Belong? The Story and Science of Invasive Species*. Vancouver, BC: Greystone Books, 2014.

Turner, Tom. "The Vindication of a Public Scholar." *Earth Island Journal*, Summer 2009.

Warren, Charles R. "Perspectives on the 'Alien' Versus 'Native' Species Debate: A Critique of Concepts, Language and Practice." *Progress in Human Geography* 31, no. 4 (2007): 427–46.

Zeidel, Robert F. *Immigrants, Progressives and Exclusion Politics: The Dillingham Commission, 1900–1927.* DeKalb: Northern Illinois University Press, 2004.

Zenderland, Leila. *Measuring Minds: Henry Herbert Goddard and the Origins of American Intelligence Testing.* New York: Cambridge University Press, 1998.

# NOTES

1: EXODUS

2    **A low fluttering cloud of butterflies** Spring Strahm, interview by author, November 5, 2018.

5    **"My goodness!"** Camille Parmesan, interview by author, January 7, 2018; "Full Interview with Camille Parmesan," University of Queensland and edX, UQx Denial 101x, YouTube, July 3, 2017; "Why I Became a Biologist: Camille Parmesan," University of Texas at Austin Environmental Science Institute, YouTube, March 6, 2007.

5    **results from her butterfly survey** Camille Parmesan, "Climate and Species' Range," *Nature* 382, no. 6594 (1996): 765.

6    **Scientists who studied everything** Camille Parmesan, "A Global Overview of Species Range Changes: Trends and Complexities; Resilience and Vulnerability," plenary speech to Species on the Move, Hobart, Tasmania, February 2016; Camille Parmesan and Mick E. Hanley, "Plants and Climate Change: Complexities and Surprises," *Annals of Botany* 116, no. 6 (2015): 849–64; Elvira S. Poloczanska et al., "Global Imprint of Climate Change on Marine Life," *Nature Climate Change* 3, no. 10 (2013): 919; I-Ching Chen et al., "Rapid Range Shifts of Species Associated with High Levels of Climate Warming," *Science* 333, no. 6045 (2011): 1024–26; Camille Parmesan, "Ecological and Evolutionary Responses to Recent Climate Change," *Annual Review of Ecology, Evolution, and Systematics* 37 (2006): 637–69; Tracie A. Scimon et al., "Upward Range Extension of Andean Anurans and Chytridiomycosis to Extreme Elevations in Response to Tropical Deglaciation," *Global Change Biology* 13, no. 1 (2007): 288–99; Craig Welch, "Half of All Species Are on the Move—And We're Feeling It," *National Geographic*, April 17, 2017.

6    ***Acropora hyacinthus*** **and** ***Acropora muricata*** Hiroya Yamano, Kaoru Sugihara, and Keiichi Nomura, "Rapid Poleward Range Expansion of Tropical Reef Corals in Response to Rising Sea Surface Temperatures," *Geophysical Research Letters* 38, no. 4 (2011).

7    **In Unalakeet, on the northwest coast** Ecological Society of America, "In a Rapidly Changing North, New Diseases Travel on the Wings of Birds," *Science*

*Daily*, December 2, 2014; Warren Richey, "Up to Cape Cod, Where No Manatee Has Gone Before," *Christian Science Monitor*, August 23, 2006.

7   **A wild exodus has begun** The terms that biologists use to refer to wild movements distinguish between types of movements depending on their perceived intent or outcome. *Range shifts* are movements that change the places where animals are generally found; *dispersals* are movements undertaken during adulthood that transfer creatures from their places birth to elsewhere, and that may or may not affect the population's distribution or range; *migrations* are purposeful to-and-fro movements, excluding those of the more circuitous and haphazard kind. In this book, I call all movements, regardless of intent or outcome, *migrations*.

8   **Every year the young saplings** Bhasha Dubey et al., "Upward Shift of Himalayan Pine in Western Himalaya, India," *Current Science*, October 2003; "Climate Change and Human Health in Tibet," Voice of America, September 12, 2015.

9   **People are on the move here** Seonaigh MacPherson et al., "Global Nomads: The Emergence of the Tibetan Diaspora (Part I)," Migration Policy Institute, September 2, 2008.

10  **Between 2008 and 2014** "Global Estimates 2015: People Displaced by Disasters," Norwegian Refugee Council and Internal Displacement Monitoring Centre, July 2015; "Global Migration Trends Factsheet," International Organization for Migration, accessed May 10, 2018; Mavroudi and Nagel, *Global Migration*; Edith M. Lederer, "UN Report: By 2030 Two-Thirds of World Will Live in Cities," Associated Press, May 18, 2016; "Over 110 Countries Join the Global Campaign to Save Productive Land," UN Convention to Combat Desertification; Robert J. Nicholls et al., "Sea-level Rise and Its Possible Impacts Given a 'Beyond 4 C World' in the Twenty-First Century," *Philosophical Transactions of the Royal Society A: Mathematical, Physical and Engineering Sciences* 369, no. 1934 (2011): 161–81.

11  **Government officials may try to estimate** Migration Policy Institute, "Mapping Fast-Changing Trends in Immigration Enforcement and Detention," Fourteenth Annual Immigration Law and Policy Conference, Georgetown University Law Center, September 25, 2017.

13  **warning of "disastrous results"** Chew, "Ending with Elton."

13  **one wrote, would not "starve gracefully"** Ehrlich, *Population Bomb*, 133.

13  **Wild species on the move** E. O. Wilson, *The Diversity of Life* (Cambridge, MA: Harvard University Press, 1992).

15  **less than 1 percent of the population** Census Organization of India, "Jain Religion Census 2011," Population Census 2011.

17    **U.S. borders had been closed** See, e.g., David Wright, Nathan Flis, and Mona Gupta, "The 'Brain Drain' of Physicians: Historical Antecedents to an Ethical Debate, c.1960–79," *Philosophy, Ethics, and Humanities in Medicine* 3, no. 1 (2008): 24; Steve Raymer, "Indian Doctors Help Fill US Health Care Needs," YaleGlobal Online, February 16, 2004; "President Lyndon B. Johnson's Remarks at the Signing of the Immigration Bill, Liberty Island, New York, October 3, 1965," Lyndon B. Johnson Presidential Library.

17    **four thousand Indian migrants to the United States** Roli Varma, "Changing Borders and Realities: Emigration of Indian Scientists and Engineers to the United States," *Perspectives on Global Development and Technology* 6, no. 4 (2007): 539–56.

22    **They glowed like beacons** Priyanka Boghani, "For Those Crossing the Mediterranean, a Higher Risk of Death," *Frontline*, October 27, 2016; Ismail Küpeli, "We Spoke to the Photographer Behind the Picture of the Drowned Syrian Boy," *Vice*, September 4, 2015; Ghulam Haqyar, interview by author, June 12, 2016.

22    **For decades the country's cruel, autocratic leaders** "Amnesty International Report 2017/18: the state of the world's human rights," Amnesty International, 2018; Patrick Kingsley, "It's Not at War, But Up to 3% of Its People Have Fled. What Is Going on in Eritrea?" *Guardian*, July 22, 2015.

22    **Mariam has a watchful way** "Mariam" and "Sophia," interviews by author, 2017. Mariam and Sophia are not their real names.

23    **Those first border crossers transformed ecosystems** Steven M. Stanley, *Earth System History*, 4th ed. (New York, Macmillan: 2015), 505–6.

24    **One of the first attempts, in 1959** Amado Araúz, "Trans-Darién Expedition 1960," Intraterra.com, archived October 27, 2009, web.archive.org/web /20091027124759/http://geocities.com/~landroverpty/trans.htm.

26    **"Whenever my son thinks about it . . ."** "Jean-Pierre" and "Mackenson," interviews by author, October 26, 201. Jean-Pierre and Mackenson are not their real names. See also Simon Nakonechny, "Pierre Recounts His Odyssey to Canada," CBC, September 26, 2017; Kate Linthicum, "Crossing the Darién Gap," *Los Angeles Times*, December 22, 2016; Lindsay Fendt, "With Olympics Over, Haitian Workers Are Leaving Brazil for the US in Big Numbers," PRI, October 4, 2016.

28    **Over 150 checkpoints, situated miles beyond** "The U.S.-Mexico Border," Migration Policy Institute, June 1, 2006, https://www.migrationpolicy.org /article/us-mexico-border. The exact number of interior checkpoints in operation is not publicly known. Simon Romero cites 170 in Romero, "Border

Patrol Takes a Rare Step in Shutting Down Inland Checkpoints," *New York Times*, March 25, 2019. See also ACLU, "The Constitution in the 100-Mile Border Zone," https://www.aclu.org/other/constitution-100-mile-border-zone.

28  **Young, strong Cesar Cuevas told me** Cesar Cuevas, interview by author, March 6, 2018; Don White, interview by author, January 8, 2018; "Bodies Found on the Border,"KVUE.com, November 7, 2016, https://www.kvue.com/video/news/local/texas-news/bodies-found-on-the-border/269-2416649.

30  **a robotics professor plotted fifteen years** Adele Peters, "Watch the Movements of Every Refugee on Earth Since the Year 2000," *Fast Company*, May 31, 2017.

30  **Over the last few years** "Global Animal Movements Based on Movebank Data (Map)," Movebank, YouTube, August 16, 2017, https://youtu.be/nUKhofr1Od8.

## 2: PANIC

32  **In late 1989 Soviet-aligned officials** "Revellers Rush on Hated Gates," *Guardian*, November 10, 1989; "February 11, 1990: Freedom for Nelson Mandela," *On This Day 1950–2005*, BBC News, http://news.bbc.co.uk/onthisday/hi/dates/stories/february/11/newsid_2539000/2539947.stm.

33  **But soon a new global bogeyman** Robert D. Kaplan, "The Coming Anarchy," *Atlantic*, February 1994.

33  **The idea of migrants as a national security threat** McLeman, *Climate and Human Migration*

34  **due to sea-level rise** McLeman, *Climate and Human Migration*, 212.

34  **One billion!** McLeman, *Climate and Human Migration,*

34  **"one of the foremost human crises . . ."** Norman Myers, "Environmental Refugees," *Population and Environment* 19, no. 2 (1997):167.

34  **They'd found that dissipating water supplies** "Water Is 'Catalyst' for Cooperation, Not Conflict, UN Chief Tells Security Council," *UN News*, June 6, 2017; T. Mitchell Aide and H. Ricardo Grau, "Globalization, Migration, and Latin American Ecosystems," *Science* 305, no. 5692 (2004): 1915–16.

34  **as a "simple stimulus-response process"** McLeman, *Climate and Human Migration*, 160.

35  **Nearly 4 million viewers tuned in** Betsy Hartmann, "Rethinking Climate Refugees and Climate Conflict: Rhetoric, Reality, and the Politics of Policy Discourse," *Journal of International Development* 22 (2010): 233–46.

35  **in the cavernous halls of the UN Security Council** McLeman, *Climate and Human Migration*, 212; "Climate Change Recognized as 'Threat Multiplier,'

UN Security Council Debates Its Impact on Peace," *UN News*, January 25, 2019.

36  **One day in early March 2011** Avi Asher-Schapiro, "The Young Men Who Started Syria's Revolution Speak About Daraa, Where It All Began," *Vice*, March 15, 2016; Michael Gunning, "Background to a Revolution," *n+1*, August 26, 2011.

37  **The war in Syria unleashed a mass exodus** Zack Beauchamp, "The Syrian Refugee Crisis, Explained in One Map," *Vox*, September 27, 2015.

37  **The flow of migrants from Syria into Europe** Anna Triandafyllidou and Thanos Maroukis, *Migrant Smuggling: Irregular Migration from Asia and Africa to Europe* (London: Palgrave Macmillan, 2012); "Mixed Migration Trends in Libya: Changing Dynamics and Protection Challenges," UNHCR, 2017.

38  **another 180,000 from the coast of Libya** "Mixed Migration Trends in Libya: Changing Dynamics and Protection Challenges," UNHCR, 2017.

38  **Filmmakers, artists, and celebrities of all ilk** Lauren Said-Moorhouse, "9 Celebrities Doing Their Part for the Refugee Crisis," CNN, December 28, 2015; Helena Smith, "Lesbos Hopes Pope's Visit Will Shine Light on Island's Refugee Role," *Guardian*, April 9, 2016; Tessa Berenson, "Susan Sarandon Is Welcoming Refugees in Greece," *Time*, December 18, 2015.

38  **Press reports immediately dubbed** Myria Georgiou and Rafal Zaborowski, "Media Coverage of the 'Refugee Crisis': A Cross-European Perspective," Council of Europe report, March 2017.

39  **Greece and Hungary had plenty** Yiannis Baboulias, "A Greek Tragedy Unfolds in Athens," *Architectural Review*, July 3, 2015; "Labour Shortages Approach Critical Level in Hungary," *Daily News Hungary*, August 15, 2016.

39  **primarily Germany but also Sweden and elsewhere** "Two Million: Germany Records Largest Influx of Immigrants in 2015," DW, March 21, 2016; Annabelle Timsit, " 'Things Could Get Very Ugly' Following Europe's Refugee Crisis," *Atlantic*, October 27, 2017; Remi Adekoya, "Why Poland's Law and Justice Party Remains So Popular," *Foreign Affairs*, November 3, 2017; "German Election: Merkel Vows to Win Back Right-Wing Voters," BBC News, September 25, 2017; "Austrian Far-Right FPÖ Draws Ire Over Refugee Internment Plan," DW, January 5, 2018; William A. Galston, "The Rise of European Populism and the Collapse of the Center-Left," Brookings Institution, March 8, 2018; "Grillo Calls for Mass Deportations (2)," *Ansa en Politics*, December 23, 2016

39  **Government agencies once dedicated to welcoming** Richard Gonzales, "America No Longer a 'Nation of Immigrants,' USCIS Says," NPR, February 22, 2018; Jennifer Rankin, " 'Do Not Come to Europe,' Donald Tusk Warns Economic Migrants," *Guardian*, March 3, 2016.

40   **scores of women showed up at police stations** Eve Hartley, "Cologne Attacks: Our Response Must Be Against Sexual Violence, Not Race, Say Feminists," *HuffPost*, January 13, 2016; Lalami, "Who Is to Blame"; Reed, "Fear and Loathing in Homer."

40   **Media outlets featured stories suggesting** Lalami, "Who Is to Blame."

41   **the German Interior Ministry released a report** Reed, "Fear and Loathing in Homer."

41   **News of the migrant-driven crime wave** Reed, "Fear and Loathing in Homer"; Eileen Sullivan, "Trump Attacks Germany's Refugee Policy, Saying US Must Avoid Europe's Immigration Problems," *New York Times*, June 18, 2018.

41   **News from Sweden a few months later** Vikas Bajaj, "Are Immigrants Causing a Swedish Crime Wave?" *New York Times*, March 2, 2017.

41   **visited Sweden to report on the situation** Ritz and Bergdahl, "People in Sweden's Alleged 'No-Go Zones.'"

42   **They called it a "no-go" zone** Ritz and Bergdahl, "People in Sweden's Alleged 'No-Go Zones"; Ami Horowitz, "Stockholm Syndrome," YouTube, December 12, 2016, https://www.youtube.com/watch?v=RqaIgeQXQgI.

42   **documentary on the migrant crisis in Sweden** Lindkvist, "Swedish Police Featured"; Dan Merica, "Trump Gets What He Wants in Florida: Campaign-Level Adulation," CNN, February 18, 2017; Rick Noack, "Trump Asked People to 'Look at What's Happening . . . in Sweden.' Here's What's Happening There," *Washington Post*, February 20, 2017.

42   **Within days, right-wing media outlets broadcast** Taylor, "Who Is Nils Bildt?"

43   **the perpetrators weren't the familiar** Marina Koren, "The Growing Fallout from the Cologne Attacks," *Atlantic*, January 11, 2016

43   **crime in Germany reached its lowest rate** "Lowest Number of Criminal Offences Since 1992," Federal Ministry of the Interior, Building, and Community, May 8, 2018

44   **The video clip of the flames** "German Police Quash Breitbart Story of Mob Setting Fire to Dortmund Church," Agence France-Presse, January 7, 2017; Reed, "Fear and Loathing in Homer."

44   **There hadn't been any crime wave in Sweden** Taylor, "Who Is Nils Bildt?"

44   **Stockholm was no "rape capital"** Ritz and Bergdahl, "People in Sweden's Alleged 'No-go Zones.'"

44   **Two of the police officers** Lindkvist, "Swedish Police Featured."

44   **Important context had been left out** "Police Close Investigation into Australian TV Crew 'Attack,'" Radio Sweden, March 1, 2016

45   **Although the number of unauthorized immigrants** Jeffrey S. Passel and D'Vera Cohn, "US Unauthorized Immigrant Total Dips to Lowest Level in a Decade," Pew Research Center, November 27, 2018; Nathan, "How the Border Patrol Faked"; U.S. Border Patrol Chief Mark Morgan and Deputy Chief Carla Provost, testimony to Senate Homeland Security and Governmental Affairs Committee, C-SPAN, November 30, 2016.

45   **bloodied bodies of two Border Patrol agents** Lam, "Border Patrol Agent"; Moore, Bever, and Miroff, "Border Patrol Agent Is Dead"; Smith, "Bannon: Killing"; "In Memoriam to Those Who Died in the Line of Duty," U.S. Customs and Border Protection, https://www.cbp.gov/about/in-memoriam /memoriam-those-who-died-line-duty; Nowrasteh, "Deaths of Border Patrol Agents."

46   **purported to reveal the security threat** Aaronson, "Trump Administration Skews"; Michael Balsamo and Colleen Long, "Trump Immigrant Crime Hotline Still Faces Hurdles, Pushback," Associated Press, February 5, 2019.

48   **Politicians and right-wing news media highlighted** "Inside ICE's Controversial Crackdown on MS-13," CBS News, November 16, 2017; "Statement from Wade on Horrific Rape in Montgomery County School," WadeKach .com, March 23, 2017, http://www.wadekach.com/blog/statement-from -wade-on-horrific-rape-in-montgomery-county-school; Zoe Chace, "Fear and Loathing in Homer and Rockville, Act Two: Loathing," *This American Life*, July 21, 2017.

48   **started to conflate migrants with criminals** Meagan Flynn, "ICE Spokesman Resigns, Citing Fabrications by Agency Chief, Sessions, About Calif. Immigrant Arrests," *Washington Post*, March 13, 2018; Mark Joseph Stern, "Trump Doesn't Need to Explain Which Immigrants He Thinks Are 'Animals,'" *Slate*, May 17, 2018; "Inside ICE's Controversial Crackdown on MS-13," CBS News, November 16, 2017.

50   **hadn't been ambushed by migrants** Between 2003 and 2017, forty of the nation's 21,000 Border Patrol agents had perished while on duty. Most died of accidents and natural causes, not by being attacked by migrants. In thirty-four of the forty deaths, the cause was either a car accident, a heart attack, or heat stress, "Border Patrol Overview," U.S. Customs and Border Protection, at https://www.cbp.gov/border-security/along-us-borders/overview; Nathan, "How the Border Patrol Faked." See also Lam, "Border Patrol Agent"; Moore, Bever, and Miroff, "Border Patrol Agent Is Dead"; Smith, "Bannon: Killing"; "In Memoriam to Those Who Died in the Line of Duty," U.S. Customs and Border Protection, https://www.cbp.gov/about/in-memoriam/memoriam -those-who-died-line-duty; Nowrasteh, "Deaths of Border Patrol Agents."

50    **the Department of Justice report** Aaronson, "Trump Administration Skews."

50    **The most gruesome and widely commented** Dan Morse, "The 'Rockville Rape Case' Erupted as National News. It Quietly Ended Friday," *Washington Post*, October 21, 2017

51    **Public health researchers in Europe** Kai Kupferschmidt, "Refugee Crisis Brings New Health Challenges," *Science*, April 22, 2016; Kirkbride, "What Are the Public Health Benefits?"; Silvia Angeletti et al., "Unusual Microorganisms and Antimicrobial Resistances in a Group of Syrian Migrants: Sentinel Surveillance Data from an Asylum Seekers Centre in Italy," *Travel Medicine and Infectious Disease* 14, no. 2 (2016): 115–22; Rein Jan Piso et al., "A Cross-Sectional Study of Colonization Rates With Methicillin-Resistant Staphylococcus aureus (MRSA) and Extended-Spectrum Beta-Lactamase (ESBL) and Carbapenemase-Producing Enterobacteriaceae in Four Swiss Refugee Centres," *PLoS One* 12, no. 1 (2017): e0170251.

51    **Fears of a migrant-driven epidemic flared** Matthew Brunwasser, "Bulgaria's Vigilante Migrant 'Hunter,'" BBC News, March 30, 2016; Kirkbride, "What Are the Public Health Benefits?"; "Thug Politics," produced by SBS (Australia), May 21, 2013. See also Helena Smith, "Golden Dawn Threatens Hospital Raids Against Immigrants in Greece," *Guardian*, June 12, 2012; Osman Dar, "Cholera in Syria: Is Europe at Risk?" *Independent*, November 2, 2015.

52    **"Tremendous infectious disease"** Philip Bump, "Donald Trump's Lengthy and Curious Defense of His Immigrant Comments, Annotated," *Washington Post*, July 6, 2015.

52    **among the most rigorously health-screened** Martin Cetron, "Refugee Crisis: Healthy Resettlement and Health Security," European Congress of Clinical Microbiology and Infectious Diseases, Amsterdam, April 12, 2016; Kirkbride, "What Are the Public Health Benefits?"

52    **"If you did that to people in the UK . . ."** Aula Abbara, interview by author, May 16, 2016.

52    **Borjas claimed to uncover** David Frum, "The Great Immigration-Data Debate," *Atlantic*, January 19, 2016.

53    **he had "nuked"** Ann Coulter, Facebook post, September 17, 2015.

53    **"the world's perhaps most effective . . ."** "Confirmation hearing on the nomination of Hon. Jeff Sessions to be Attorney General of the United States," Committee on the Judiciary, U.S. Senate, January 10–11, 2017 (Washington, D.C.: U.S. Government Printing Office); "How Sessions and Miller Inflamed Anti-Immigrant Passions from the Fringe," *New York Times*, June 19, 2018; Philip Bump, "A Reporter Pressed the White House for Data. That's When Things Got Tense," *Washington Post*, August 2, 2017.

54  **left out a potentially confounding factor** Michael Clemens, "What the Mariel Boatlift of Cuban Refugees Can Teach Us About the Economics of Immigration," Center for Global Development, May 22, 2017.

54  **cost the United States "billions"** "Fact Check: Trump's First Address to Congress," *New York Times*, February 28, 2017.

54  **the economic benefits contributed** Julie Hirschfield Davis and Somini Sengupta, "Trump Administration Rejects Study Showing Positive Impact of Refugees," *New York Times*, September 18, 2017; "Fact Check: Trump's First Address to Congress," *New York Times*, February 28, 2017.

55  **overrepresented in federal crime statistics** Salvador Rizzo, "Questions Raised About Study That Links Undocumented Immigrants to Higher Crime," *Washington Post*, March 21, 2018; Alex Nowrasteh, "The Fatal Flaw in John R. Lott Jr.'s Study of Illegal Immigrant Crime in Arizona," Cato Institute, February 5, 2018; John R. Lott, "Undocumented Immigrants, US Citizens, and Convicted Criminals in Arizona," 2018; Jonathan Hanen, Greater Towson Republican Club, Towson, Md., January 16, 2018. Biographical details from Jonathan Hanen's public profile are on LinkedIn at https://www.linkedin.com/in/jonathan-hanen-89a93715.

56  **infiltrated the cultural conversation** Reed, "Fear and Loathing in Homer."

57  **a picture of migrants as a global threat** Eduardo Porter and Karl Russell, "Immigration Myths and Global Realities," *New York Times*, June 20, 2018; Richard Wike, Bruce Stokes, and Katie Simmons, "Europeans Fear Wave of Refugees Will Mean More Terrorism, Fewer Jobs," Pew Research Center, July 11, 2016; Salvador Rizzo, "Questions Raised About Study That Links Undocumented Immigrants to Higher Crime," *Washington Post*, March 21, 2018.

57  **The president described a 2018 caravan** Jeremy W. Peters, "How Trump-fed Conspiracy Theories About Migrant Caravan Intersect with Deadly Hatred," *New York Times*, October 29, 2018.

58  **"Why are we having all these people from shithole countries . . ."** The White House later denied the comments, but the *New York Times* stood by its reporting. Michael D. Shear and Julie Hirschfeld Davis, "Stoking Fears, Trump Defied Bureaucracy to Advance Immigration Agenda," *New York Times*, December 23, 2017; Josh Dawsey, "Trump Derides Protections for Immigrants from 'Shithole' Countries," *Washington Post*, January 12, 2018.

58  **Then the Brazilian economy tanked** Emily Gogolak, "Haitian Migrants Turn Toward Brazil," *New Yorker*, August 20, 2014; Olivier Laurent, "These Haitian Refugees Are Stranded at the U.S.-Mexico Border," *Time*, February 20, 2017.

59  he'd found that Haiti had "made significant progress" Rivas, "DHS Ignored Its Own."

59  who'd established homes and businesses Rhina Guidos, "Study Says Doing Away With Immigration Program Would Harm Economy," *National Catholic Reporter*, July 27, 2017.

59  Emmanuel Louis, a lawyer from Port-au-Prince "Emmanuel Louis," interview by author, October 24, 2017. Emmanuel Louis is not his real name.

59  Community workers across the country Gabeau interview.

60  Jean-Pierre's family barely escaped summary deportation "Jean-Pierre" interview.

60  to build its argument for Haitians' eviction Alicia A. Caldwell, "Haitians Under the Microscope," Associated Press, May 9, 2017.

61  According to a State Department travel advisory Haiti Travel Warning, September 12, 2017, U.S. Passports and International Travel, U.S. Department of State.

61  ignored the findings Rivas, "DHS Ignored Its Own."

61  "If we don't do something about the border . . ." Samuel Granados et al., "Raising Barriers: A New Age of Walls: Episode 1," *Washington Post*, October 12, 2016.

## 3: LINNAEUS'S LOATHSOME HARLOTRY

63  As a boy, Linnaeus spent his days Blunt, *Linnaeus*, 14.

64  Prince of Flowers Koerner, *Linnaeus*, 84.

65  overflowing with breathless tales Broberg, "*Homo sapiens*," 185–86, 191; Curran, *Anatomy of Blackness*, 106–9, 144.

65  eighteenth-century travel writers underlined Bendyshe, "History of Anthropology"; Schiebinger, "Taxonomy for Human Beings"; Fausto-Sterling, "Gender, Race, and Nation."

65  these tales were mostly cobbled together Schmidt, *Inventing Exoticism*, 55.

66  Voltaire's description of cave-dwelling peoples Cat Bohannon, "The Curious Case of the London Troglodyte," *Lapham's Quarterly*, June 15, 2013.

66  Kioping's description of his encounter Christina Skott, "Linnaeus and the Troglodyte: Early European Encounters with the Malay World and the Natural History of Man," *Indonesia and the Malay World* 42, no. 123 (2014): 141–69; Maya Wei-Haas, "The Hunt for the Ancient 'Hobbit's' Modern Relatives," *National Geographic*, August 2, 2018; Brian Handwerk, "Saint Nicholas to Santa: The Surprising Origins of Mr. Claus," *National Geographic*, November 29, 2017.

67  certain aspects of foreign people Jablonski, *Living Color*; Fausto-Sterling, "Gender, Race, and Nation"; Schmidt, *Inventing Exoticism*, 1–33.

68  a distinctly different experience of human diversity Sussman, *Myth of Race.*

68  Depictions of and stories about these strange foreign others Schmidt, *Inventing Exoticism*, 1–33; Blunt, *Linnaeus*; Fausto-Sterling, "Gender, Race, and Nation."

69  Debate swirled among intellectuals and elites Sloan, "Gaze of Natural History."

70  Linnaeus had little direct knowledge Blunt, *Linnaeus*; Koerner, *Linnaeus*, 57.

71  His benefactors were impressed Blunt, *Linnaeus*, 96–99; Koerner, *Linnaeus*, 16.

72  Linnaeus started writing his groundbreaking taxonomy Jonathan Marks, "Long Shadow of Linnaeus's Human Taxonomy," *Nature*, May 3, 2007.

72  When it came to describing humans Bendyshe, "History of Anthropology."

73  He described botanical marriages Schiebinger, *Nature's Body*, 21.

73  "Every animal feels the sexual urge . . ." Blunt, *Linnaeus*, 33.

73  Critics decried it Blunt, *Linnaeus*, 121.

74  Linnaeus's rival Richard Conniff, "Buffon: Forgotten, Yes. But Happy Birthday Anyway," *New York Times*, January 2, 2008.

74  Buffon saw it as mutable and dynamic Paul L. Farber, "Buffon and the Concept of Species," *Journal of the History of Biology* 5, no. 2 (1972): 259–84, www.jstor.org/stable/4330577; "Heraclitus," *Stanford Encyclopedia of Philosophy*, June 23, 2015, https://plato.stanford.edu/entries/heraclitus/.

75  For Buffon, albinism among Africans Marks, *Human Biodiversity*, 120; Curran, *Anatomy of Blackness*, 88, 106.

76  Perhaps, he speculated, sometime in our deep past Sloan, "Gaze of Natural History."

76  The idea that weather patterns Frederick Foster and Mark Collard, "A Reassessment of Bergmann's Rule in Modern Humans," *PLOS One* 8, no. 8 (2013): e72269; Ann Gibbons, "How Europeans Evolved White Skin," *Science*, April 2, 2015; Angela M. Hancock et al., "Adaptations to Climate in Candidate Genes for Common Metabolic Disorders," *PLOS Genetics* 4, no. 2 (2008): e32; Maria A. Serrat, Donna King, and C. Owen Lovejoy, "Temperature Regulates Limb Length in Homeotherms by Directly Modulating Cartilage Growth," *Proceedings of the National Academy of Sciences* 105, no. 49 (2008): 19348–53.

79  The degenerative effects of migration Sloan, "Gaze of Natural History."

80 **Top scientific societies** Jablonski, *Living Color*; Lee Alan Dugatkin, "Thomas Jefferson Defends America with a Moose," *Slate*, September 12, 2012; Ernst Mayr, *The Growth of Biological Thought* (Cambridge, MA: Harvard University Press, 1981), 330; "Buffon, Georges-Louis Leclerc, Comte De," *Complete Dictionary of Scientific Biography* (New York: Charles Scribner's Sons, 2008), https://www.encyclopedia.com/people/science-and-technology/geology-and-oceanography-biographies/georges-louis-leclerc-buffon-comte-de.

80 **Linnaeus was not impressed** Koerner, *Linnaeus*, 28.

80 **made the naming of living things uniform and universal** Broberg, "Anthropomorpha," 95; Bendyshe, "History of Anthropology."

80 **"Take a bird or a lizard or a flower . . ."** Anne Fadiman, *At Large and at Small: Familiar Essays* (New York: Farrar, Straus and Giroux, 2008), 19; Richard Holmes, *The Age of Wonder: How the Romantic Generation Discovered the Beauty and Terror of Science* (New York: Knopf, 2009), 49.

81 **Luminaries and royal patrons** Smethurst, *Travel Writing and Natural World*; Blunt, *Linnaeus*, 153–58.

81 **Linnaeus discounted even the most obvious** Nancy J. Jacobs, "Africa, Europe, and the Birds Between Them," in James Beattie, Edward Melillo, and Emily O'Gorman, *Eco-cultural Networks and the British Empire: New Views on Environmental History* (New York: Bloomsbury Academic, 2015).

82 **In the sixteenth century** See, e.g., Andrew J. Lewis, *A Democracy of Facts: Natural History in the Early Republic* (Philadelphia: University of Pennsylvania Press, 2011).

82 **the alternative idea that birds annually traveled thousands of miles** Dingle, *Migration*; Ron Cherry, "Insects and Divine Intervention," *American Entomologist* 61, no. 2 (2015): 81–84, https://doi.org/10.1093/ae/tmv001.

82 **For Linnaeus, there'd been only a single dispersal** Jorge Crisci et al., *Historical Biogeography: An Introduction* (Cambridge, MA: Harvard University Press, 2009), 30; Bendyshe, "History of Anthropology."

83 **"As he is rather eloquent . . ."** Lisbet Koerner, "Purposes of Linnaean Travel: A Preliminary Research Report," in David Philip Miller and Peter Hanns Reill, eds., *Visions of Empire: Voyages, Botany, and Representations of Nature* (New York: Cambridge University Press, 2011), 119.

83 **He named a plant after the comte** Koerner, *Linnaeus*, 28; Richard Conniff, "Forgotten, Yes. But Happy Birthday Anyway," *New York Times*, December 30, 2007.

83 **With his tenth and most authoritative edition** Smethurst, *Travel Writing and Natural World*.

85    van Leeuwenhoek's inquiries led him to believe Curran, *Anatomy of Blackness.*

85    Sexual anatomy fascinated Linnaeus Jonathan Marks, interview by author, September 5, 2017; Blunt, *Linnaeus.*

86    It was known as the "Hottentot apron" Schiebinger, *Nature's Body*, 170.

86    At first, European travelers speculated Rachel Holmes, *African Queen: The Real Life of the Hottentot Venus* (New York: Random House, 2009); Curran, *Anatomy of Blackness*, 109; Switek, "Tragedy of Baartman."

86    Certain humans, he said, were a separate species altogether Koerner, *Linnaeus*, 57; Bendyshe, "History of Anthropology."

87    Linnaeus speculated, privately "A flying thought," Linnaeus wrote, "could give the idea that some woman had mixed with the troglodytes and that [it is from this that] the Hottentot take their origin," Broberg, "Anthropomorpha," 95; Marks, *Human Biodiversity*, 50.

87    Linnaeus proclaimed natural history's independence Gould, *Flamingo's Smile.*

88    Linnaeus's "rapid historical triumph" over Buffon Phillip R. Sloan, "The Buffon-Linnaeus Controversy," *Isis* 67, no. 3 (1976): 356–75; Curran, *Anatomy of Blackness*, 169.

88    In 1774 Louis XV ordered Blunt, *Linnaeus.*

89    The most explosive claim in Linnaean taxonomy Blunt, *Linnaeus*; Broberg, "*Homo sapiens*," 178.

90    First he tried to buy the girl Broberg, "*Homo sapiens*," 185–86.

90    European scientists continued to be foiled Gould, *Flamingo's Smile*; Schiebinger, "Taxonomy for Human Beings."

90    Europe's most famous scientists flocked to view Switek, "Tragedy of Baartman."

90    Cuvier arranged for a commission Schiebinger, *Nature's Body*, 170; Schiebinger, "Taxonomy for Human Beings."

91    He didn't find anything like Gould, *Flamingo's Smile.*

91    Still, absence of evidence Gould, *Flamingo's Smile*; Schiebinger, "Taxonomy for Human Beings"; Clifton C. Crais and Pamela Scully, *Sara Baartman and the Hottentot Venus: A Ghost Story and a Biography* (Princeton, NJ: Princeton University Press, 2009).

91    Linnaean taxonomy formed the basis Koerner, *Linnaeus*; Broberg, "Anthropomorpha," 95.

92    The eighteenth-century naturalist Pierre-Louis Moreau William B. Cohen, *The French Encounter with Africans: White Response to Blacks, 1530–1880* (Bloomington: Indiana University Press, 2003), 86.

## 4: THE DEADLY HYBRID

95 **At the same time, people from** Zeidel, *Immigrants, Progressives.*

95 **The newcomers took jobs peddling** Tyler Anbinder, *Five Points: The 19th-Century New York City Neighborhood That Invented Tap Dance, Stole Elections, and Became the World's Most Notorious Slum* (New York: Plume, 2001), 43.

95 **The lifestyle of the old New York elites** Sussman, *Myth of Race.*

96 **Osborn and Grant belonged** Spiro, *Defending the Master Race,* 25.

97 **"Viewed zoologically . . ."** James Lander, *Lincoln and Darwin: Shared Visions of Race, Science, and Religion* (Carbondale: Southern Illinois University Press, 2010), 81.

97 **"pure racial types"** Brian Wallis, "Black Bodies, White Science: Louis Agassiz's Slave Daguerreotypes," *American Art* 9, no. 2 (1995): 39–61.

97 **Naturalists had become so convinced** Marks, *Human Biodiversity,* 125; Davenport et al., *Eugenics in Race and State.*

98 **The political and economic value** Massin, "From Virchow to Fischer."

99 **Charles Darwin had purposely omitted** E. J. Browne, *Charles Darwin: The Power of Place* (New York: Knopf, 2002), 42; Edward Lurie, "Louis Agassiz and the Idea of Evolution," *Victorian Studies* 3, no. 1 (1959): 87–108.

99 **For Darwin, differences between peoples** Darwin, *Descent of Man,* 202–3.

99 **But as the race scientists grew more confident** Darwin, *Descent of Man,* xxxviii.

100 **By the time he published *The Descent of Man*** Massin, "From Virchow to Fischer"; Darwin, *Descent of Man,* xxxiv; Peter J. Bowler, *Evolution: The History of an Idea* (Berkeley: University of California Press, 2003), 224–25.

100 **"Darwin's greatest unread book"** Darwin, *Descent of Man,* lv.

100 **And so science populizers such as Osborn and Grant** Spiro, *Defending the Master Race,* 46; Mitch Keller, "The Scandal at the Zoo," *New York Times,* August 6, 2006; Pierpont, "Measure of America."

102 **Weismann's experiments did not** Sussman, *Myth of Race*; Herbert Eugene Walter, *Genetics: An Introduction to the Study of Heredity* (New York: Macmillan, 1913); Daniel J. Kevles, *In the Name of Eugenics: Genetics and the Uses of Human Heredity* (New York: Knopf, 1985); Davenport, *Heredity in Relation,* 24; Nathaniel Comfort, *The Science of Human Perfection: How Genes Became the Heart of American Medicine* (New Haven, CT: Yale University Press, 2012), 44; Mukherjee, *Gene,* 64.

104 **"the most stable form of matter . . ."** Osborn, "Poor Nordic!"

105 **Scientific concerns about sexual relations** Spiro, *Defending the Master Race,* 92–94; Provine, "Geneticists and the Biology"; Nancy Stepan, "Biological Degeneration: Races and Proper Places," in Chamberlin and Gilman, *Degeneration.*

106 **The precise outcome of racial hybridization** Charles B. Davenport, "The Effects of Race Intermingling," *Proceedings of the American Philosophical Society* 56, no. 4 (1917): 364–68; Spiro, *Defending the Master Race*, 95.

106 **"rapidly become darker in pigmentation . . ."** Davenport, *Heredity in Relation*, 219.

106 **"absolute ruin"** Black, *War Against the Weak*.

106 **"Miscegenation," Grant wrote** Spiro, *Defending the Master Race*, 152.

107 **Congress had closed U.S. borders** Zeidel, *Immigrants, Progressives*, 113; Spiro, *Defending the Master Race*, 46; Sussman, *Myth of Race*, 61.

107 **While scientific elites detailed the biological menace** Howard Markel and Alexandra Minna Stern, "The Foreignness of Germs: The Persistent Association of Immigrants and Disease in American Society," *Milbank Quarterly* 80, no. 4 (2002): 757–88; Harvey Levenstein, "The American Response to Italian Food, 1880–1930," *Food and Foodways* 1, nos. 1–2 (1985): 1–23.

108 **President Roosevelt attended on opening night** Charles Hirschman, "America's Melting Pot Reconsidered," *Annual Review of Sociology* 9, no. 1 (1983): 397–423; "President Sees New Play," *New York Times*, October 6, 1908; "Roosevelt Criticises Play," *New York Times*, October 10, 1908.

108 **"the more the merrier"** Zeidel, *Immigrants, Progressives*, 35.

109 **Boas's study found that** Benton-Cohen, *Inventing the Immigration Problem*; Zeidel, *Immigrants, Progressives*, 71–78, 100.

111 **Grant privately scoffed** Spiro, *Defending the Master Race*, 199.

111 **"keep out a great mass of worthless Jews . . ."** Zeidel, *Immigrants, Progressives*, 125.

111 **"study it"** Tamsen Wolff, *Mendel's Theatre: Heredity, Eugenics, and Early Twentieth-Century American Drama* (New York: Palgrave, 2009)

111 **"a hybrid race of people as worthless . . ."** Spiro, *Defending the Master Race*, 174.

111 **At universities across the country** Sussman, *Myth of Race*; Black, *War Against the Weak*.

112 **Anti-German propaganda and anxieties** Zenderland, *Measuring Minds*, 276; Harry H. Laughlin, "Nativity of Institutional Inmates," in Davenport et al., *Eugenics in Race and State*; "Says Insane Aliens Stream in Steadily," *New York Times*, June 5, 1924; Edwin Fuller Torrey and Judy Miller, *The Invisible Plague: The Rise of Mental Illness from 1750 to the Present* (New Brunswick, NJ: Rutgers University Press, 2001); Leon Kamin, *The Science and Politics of IQ* (Mahwah, NJ: Lawrence Erlbaum Associates, Psychology Press, 1974), 15–32.

113 **Methodological biases accounted for the findings** Zenderland, *Measuring Minds*, 286.

113  **At Ellis Island, officials administered** Allan V. Horwitz and Gerald N. Grob, "The Checkered History of American Psychiatric Epidemiology," *Milbank Quarterly*, December 2011.

113  **Subspecies theory predicted that hybrids** Franz Boas, "The Half-Blood Indian: An Anthropometric Study," *Popular Science Monthly*, October 1894; Massin, "From Virchow to Fischer"; Herman Lundborg, "Hybrid Types of the Human Race," *Journal of Heredity* (June 1921).

114  **innuendo and speculation** Charles B. Davenport, "The Effects of Race Intermingling," *Proceedings of the American Philosophical Society* 56, no. 4 (1917): 364–68; Nancy Stepan, "Biological Degeneration: Races and Proper Places," in Chamberlin and Gilman, *Degeneration*; Black, *War Against the Weak*.

116  **Osborn was in the midst of organizing** Anderson, "Racial Hybridity, Physical Anthropology"; Anderson, "Hybridity, Race, and Science"; Frederick Hoffman, "Race Amalgamation in Hawaii," in Davenport et al., *Eugenics in Race and State*, 90–108; "Museum History: A Timeline," American Museum of Natural History, https://www.amnh.org/about/timeline-history.

116  **"We are engaged in a serious struggle . . ."** Osborn, "Poor Nordic!"; "Tracing Parentage by Eugenic Tests," *New York Times*, September 23, 1921.

117  **The Census Bureau provided several diagrams** Gelb, Allen, Futterman, and Mehler, "Rewriting Mental Testing"; Spiro, *Defending the Master Race.*

117  **For a week, the gathered attendees** Leonard Darwin, "The Field of Eugenic Reform," in Davenport et al., *Eugenics in Race and State*; "Tracing Parentage by Eugenic Tests," *New York Times*, September 23, 1921

118  **"I'm head over heels in the Polynesian problem"** Anderson, "Racial Hybridity, Physical Anthropology"; Laughlin, *Second International Exhibition of Eugenics.*

118  **A colleague appeared at the conference** L. C. Dunn, "Some Results of Race Mixture in Hawaii," and Maurice Fishberg, "Intermarriage Between Jews and Christians," both in Davenport et al., *Eugenics in Race and State*, 109–24.

120  **"nourish and develop a strong and healthy race instinct"** Jon Alfred Mjøen, "Harmonic and Disharmonic Racecrossings," in *Scientific Papers of the Second International Congress of Eugenics Held at the American Museum of Natural History, New York, September 22–28, 1921* (Baltimore: Williams & Wilkins, 1923), vol. 2.

120  **When the week came to a close** Gelb et al., "Rewriting Mental Testing History."

122  **Laws that temporarily and partially restricted immigration** Spiro, *Defending the Master Race*, 216, 221, 225; Gelb et al., "Rewriting Mental Testing History"; "1890 Census Urged as Immigrant Base," *New York Times*, January 7, 1924;

Kenneth M. Ludmerer, *Genetics and American Society: A Historical Appraisal* (Baltimore: Johns Hopkins University Press, 1972).

122 **And yet as his research progressed** Anderson, "Racial Hybridity, Physical Anthropology"; Anderson, "Hybridity, Race, and Science."

124 **bad teeth** Shapiro, *Pitcairn Islanders*; "Dr. Harry L. Shapiro, Anthropologist, Dies at 87," *New York Times*, January 9, 1990.

124 **"Physically there is little to choose . . ."** Provine, "Geneticists and the Biology."

125 **"Man emerges as a dynamic organism"** Jonathan Marks, interview by author, September 5, 2017; Anderson, "Racial Hybridity, Physical Anthropology."

126 **"an integral factor in the history of human civilization"** Frank Spencer, "Harry Lionel Shapiro: March 19, 1902–January 7, 1990," in National Academy of Sciences, *Biographical Memoirs* (Washington, D.C.: National Academies Press, 1996), vol. 70, https://doi.org/10.17226/5406.

126 **The flow of immigrants into the United States** Mavroudi and Nagel, *Global Migration*; Benton-Cohen, *Inventing the Immigration Problem*; Zeidel, *Immigrants, Progressives*, 146.

126 **"The book is my bible"** Spiro, *Defending the Master Race*, 357.

126 **"We must ignore the tears of sobbing sentimentalists . . ."** Spiro, *Defending the Master Race*, 370.

126 **A few months later an ocean liner** Dara Lind, "How America's Rejection of Jews Fleeing Nazi Germany Haunts Our Refugee Policy Today," *Vox*, January 27, 2017; Ishaan Tharoor, "What Americans Thought of Jewish Refugees on the Eve of World War II," *Washington Post*, November 17, 2015.

5: THE SUICIDAL ZOMBIE MIGRANT

128 **The expedition provided a valuable opportunity** Crowcroft, *Elton's Ecologists*, 4.

129 **Elton did not happen upon his chance** Anker, *Imperial Ecology*; Nils Christian Stenseth, "On Evolutionary Ecology and the Red Queen," YouTube, January 12, 2017, https://www.youtube.com/watch?v=Rwc9WI_a2Nw.

131 **The mysterious cycling of populations** Mark A. Hixon et al., "Population Regulation: Historical Context and Contemporary Challenges of Open vs. Closed Systems," *Ecology* 83, no. 6 (2002): 1490–508; Chitty, *Do Lemmings Commit Suicide?*; Anker, *Imperial Ecology*.

132 **reports of lemmings gathering together** Duppa Crotch, "The Migration of the Lemming," *Nature* 45, no. 1157 (1891); Crotch, "Further Remarks on the Lemming."

132 **In 1888 a mass of lemmings formed** Stenseth and Ims, *Biology of Lemmings*; Chitty, *Do Lemmings Commit Suicide?*; Anker, *Imperial Ecology*; Crotch, "Further Remarks on the Lemming."

133 **Collett's stories struck Elton differently** Crowcroft, *Elton's Ecologists*, 4; Bashford, *Global Population.*

134 **Elton wrote up his novel spin on Collett's findings** Lindström, "From Arctic Lemmings"; Peder Anker, interview by author, February 7, 2018; Elton, "Periodic Fluctuations."

134 **"The phenomenon," he explained** Elton, "Periodic Fluctuations."

134 **Elton had discovered the mysterious factor X** Lindström, "From Arctic Lemmings."

135 **Biologists discovered manifestations of this secret drive** Marston Bates, *The Nature of Natural History*, vol. 1138 (Princeton, NJ: Princeton University Press, 2014); Ramsden and Wilson, "Suicidal Animal."

136 **For Elton, as for Linnaeus, the conviction** Chew, "Ending with Elton"; Charles S. Elton, *Animal Ecology* (Chicago: University of Chicago Press, 2001).

137 **This conception of the past conformed** Chew, "Ending with Elton."

138 **According to Gause's Law** Georgii Frantsevich Gause, "Experimental Studies on the Struggle for Existence: I. Mixed Population of Two Species of Yeast," *Journal of Experimental Biology* 9, no. 4 (1932): 389–402.

138 **Years of experimentation and mathematical modeling** Garrett Hardin, "The Competitive Exclusion Principle," *Science* 131, no. 3409 (1960): 1292–97.

138 **In the belief that nature was in essence "filled up"** Chew, "Ending with Elton"; Peter Coates, *American Perceptions of Immigrant and Invasive Species: Strangers on the Land* (Berkeley: University of California Press, 2007).

140 **In Germany, people purged plants deemed foreign** Joachim Wolschke-Bulmahn and Gert Groening, "The Ideology of the Nature Garden: Nationalistic Trends in Garden Design in Germany During the Early Twentieth Century," *Journal of Garden History* 12, no. 1 (1992): 73–78; Daniel Simberloff, "Confronting Introduced Species: A Form of Xenophobia?" *Biological Invasions* 5 (2003): 179–92; Spiro, *Defending the Master Race*, 379.

140 **Elton did not explicitly extend the implications** Thomas Robertson, "Total War and the Total Environment: Fairfield Osborn, William Vogt, and the Birth of Global Ecology," *Environmental History* 17, no. 2 (April 2012): 336–64; Ramsden and Wilson, "Suicidal Animal"; Anker, *Imperial Ecology.*

141 **Elton's ideas shed "considerable light . . ."** Chew, "Ending with Elton."

141 **By the 1930s, the popularity of eugenics** Pierpont, "Measure of America."

141 **But scientists did not abandon their suspicions** Chew, "Ending with Elton."

142 **Many of the billions of birds that migrate** Dingle, *Migration*, 48; Kessler, "Most Extreme Migration?"; L. R. Taylor, "The Four Kinds of Migration," in W. Danthanarayana, ed., *Insect Flight: Proceedings in Life Sciences* (Berlin: Springer, 1986): 265–80.

143 **The ornithologist David Lack had a theory** Ted R. Anderson, *The Life of David Lack: Father of Evolutionary Ecology* (New York: Oxford University Press, 2013); "Radar 'Bugs' Found to Be—Just Bugs," *New York Times*, April 4, 1949; Chew, "Ending with Elton"; David Lack and G. C. Varley, "Detection of Birds by Radar," *Nature*, October 13, 1945; "Messerschmitt Bf 109," MilitaryFactory .com, https://www.militaryfactory.com/aircraft/detail.asp?aircraft_id=83; I. O. Buss, "Bird Detection by Radar," *Auk* 63 (1946): 315–18; David Clarke, "Radar Angels," *Fortean Times* 195 (2005), https://drdavidclarke.co.uk/secret -files/radar-angels

146 **"practically in league with the Nazis"** Thompson, *Where Do Camels Belong?*, 39.

146 **the arrival of Asian chestnut trees** Mark A. Davis, Ken Thompson, and J. Philip Grime, "Charles S. Elton and the Dissociation of Invasion Ecology from the Rest of Ecology," *Diversity and Distributions* 7 (2001): 97–102; Gintarė Skyrienė and Algimantas Paulauskas, "Distribution of Invasive Muskrats (*Ondatra zibethicus*) and Impact on Ecosystem," *Ekologija* 58, no. 3 (2012); Elton, *Ecology of Invasions*, 21–27; Harold A. Mooney and Elsa E. Cleland, "The Evolutionary Impact of Invasive Species," *Proceedings of the National Academy of Sciences* 98, no. 10 (2001): 5446–51.

146 **In his postwar books, radio addresses, and papers** Chew, "Ending with Elton"; Thompson, *Where Do Camels Belong?* 39.

147 **Even if newly arrived species seem benign** Daniel Simberloff, foreword to Elton, *Ecology of Invasions*, xiii; Thompson, *Where Do Camels Belong?* 39.

147 **To depict species on the move as "invaders"** Mark A. Davis et al., "Don't Judge Species on Their Origins," *Nature* 474, no. 7350 (2011): 153–54; Matthew K. Chew, "Indigene Versus Alien in the Arab Spring: A View Through the Lens of Invasion Biology," in Uzi Rabi and Abdelilah Bouasria, eds., *Lost in Translation: New Paradigms for the Arab Spring* (Eastbourne, UK: Sussex Academic Press, 2017).

148 **Elton delivered his warnings about invasive species** Matthew K. Chew, "A Picture Worth Forty-One Words: Charles Elton, Introduced Species and the 1936 Admiralty Map of British Empire Shipping," *Journal of Transport History* 35, no. 2 (2014): 225–35.

148 **"one of the central scientific books of our century"** David Quammen, back cover blurb to Elton, *Ecology of Invasions*.

148 **Produced by Walt Disney studios** Jack Jungmeyer, "Filming a 'Wilderness,'" *New York Times*, August 3, 1958; *Cruel Camera: Animals in Movies*, documentary film, Fifth Estate program, CBC Television, May 5, 1982.

150 **Elton is remembered today as the "founding father"** Richard Southwood and J. R. Clarke, "Charles Sutherland Elton: 29 March 1900–1 May," *Biographical Memoirs of Fellow of the Royal Society*, November 1, 1999; Chitty, *Do Lemmings Commit Suicide?*

151 **The truth about lemmings emerged** Tim Coulson and Aurelio Malo, "Case of the Absent Lemmings," *Nature*, November 2008; Chitty, *Do Lemmings Commit Suicide?*; Nils Christian Stenseth, interview by author, February 9, 2018.

151 **During snowy years, unknown to anyone** Nicholls, "Truth About Norwegian Lemmings.

151 **"cock-and-bull stories from Norwegian sailors"** Anker, *Imperial Ecology*.

151 **The lemming suicide march had been staged** *Cruel Camera: Animals in Movies*, documentary film, Fifth Estate program, CBC Television May 5, 1982; Nicholls, "Truth About Norwegian Lemmings."

152 *White Wilderness* **infiltrated the public mind for decades** Jim Korkis, "Walt and the True-Life Adventures," Walt Disney Family Museum, February 9, 2012.

152 **The lemmings' macabre migration captivated the nation** Stenseth and Ims, *Biology of Lemmings*; *Columbia Anthology of British Poetry* (New York: Columbia University Press, 2010), 808.

153 **During the war, millions had marched to their deaths** Ramsden and Wilson, "Suicidal Animal"; Robertson, *Malthusian Moment*.

6: MALTHUS'S HIDEOUS BLASPHEMY

154 **The cull transformed the plateau** Frederick Andrew Ford, *Modeling the Environment*, 2nd ed. (Washington, D.C.: Island Press, 2009), 267–72; D. R. Klein, "The Introduction, Increase, and Crash of Reindeer on St. Matthew Island," *Journal of Wildlife Management* 32 (1968): 3S0367; Ned Rozell, "When Reindeer Paradise Turned to Purgatory," University of Alaska Fairbanks Geophysical Institute, August 9, 2012.

156 **the "demographic transition"** Jeremy Greenwood and Ananth Seshadri, "The US Demographic Transition," *American Economic Review* 92, no. 2 (2002): 153–59; Friedrich Engels, "Outlines of a Critique of Political Economy," *Deutsch-Französische Jahrbücher* 1 (1844).

158 **"remarkable lack of wanderlust"** Paul R. Ehrlich and Ilkka Hanski, *On the Wings of Checkerspots: A Model System for Population Biology* (New York: Oxford University Press, 2004).

159 **"The streets seemed alive with people . . ."** Ehrlich, *Population Bomb*.

159 **For Ehrlich** Robertson, *Malthusian Moment*.

160 **In fact, the crowds and environmental damage** Mann, "Book That Incited"; Robertson, *Malthusian Moment*; Jennifer Crook, "War in Kashmir and Its Effect on the Environment," *Inventory of Conflict and Environment*, April 16, 1998, http://mandalaprojects.com/ice/ice-cases/kashmiri.htm.

160 **Ehrlich's crusade to arrest the growth** Turner, "Vindication"; Ramsden, "Confronting the Stigma"; Gutiérrez, *Fertile Matters*; "History," California Air Resources Board, https://ww2.arb.ca.gov/about/history; Rian Dundon, "Photos: L.A.'s Mid-Century Smog Was So Bad, People Thought It Was a Gas Attack," Timeline, May 23, 2018, https://timeline.com/la-smog-pollution -4ca4bc0cc95d.

161 **In his influential study** Ramsden and Adams, "Escaping the Laboratory."

162 **Ehrlich cited what he considered their foregone conclusions** Ramsden and Adams, "Escaping the Laboratory"; Ramsden, "Confronting the Stigma."

162 **Scientists started to refer to human population growth** Desrochers and Hoffbauer, "Postwar Intellectual Roots."

163 **Congress passed the Hart-Celler Act in 1965** Gabriel Chin and Rose Cuison Villazor, eds., *The Immigration and Nationality Act of 1965: Legislating a New America* (New York: Cambridge University Press, 2015).

164 **The population bomb would not be contained** Josh Zeitz, "The 1965 Law That Gave the Republican Party Its Race Problem," *Politico*, August 20, 2016; Paul R. Ehrlich and John P. Holdren, "Impact of Population Growth," *Science*, March 26, 1971.

164 **Population growth was a problem that implicated everyone** "The Population Bomb?" *New York Times*, May 31, 2015; Turner, "Vindication"; Ehrlich, *Population Bomb*, 130, 151–52.

165 **Ehrlich was not an overt racist** Ramsden, "Confronting the Stigma"; Edward B. Fiske, "Argument by Overkill," *New York Times*, October 1, 1977.

166 **a theory called "r/K selection" consumed population biologists** David Reznick, Michael J. Bryant, and Farrah Bashey, "r-and K-selection Revisited: The Role of Population Regulation in Life-History Evolution," *Ecology* 83, no. 6 (2002): 1509–20.

166 **Rushton would explicitly apply r/K selection theory to human racial groups** J. Philippe Rushton, "Race, Evolution, Behavior (abridged version)," Port Huron, MI: Charles Darwin Research Institute, 2000.

167 **Ehrlich's bias was not overt, but** Ehrlich, *Population Bomb*, 80–84; Ehrlich, interview by WOI-TV.

167 **"Countries are divided rather neatly into two groups . . ."** Ehrlich, *Population Bomb*, 7.

167 **Ehrlich's fellow neo-Malthusian scientists** Kingsley Davis, "The Migrations of Human Populations," *Scientific American*, September 1974; Ehrlich, *Population Bomb*; Horowitz, *Anxieties of Affluence.*

167 **"I know this all sounds very callous . . ."** Ehrlich, *Population Bomb*, 151–52.

168 **Then in early 1970, Johnny Carson called** MediaVillage, "History's Moment in Media: Johnny Carson Became NBC's Late-Night Star," A+E Networks, May 22, 2018, https://www.mediavillage.com/article /HISTORYS-Moment-in-Media-Johnny-Carson-Became-NBCs-Late -Night-Star/; Mark Malkoff, *The Carson Podcast with Guest Dr. Paul Ehrlich*, April 12, 2018; "The Population Bomb?" *New York Times*, May 31, 2015; Robertson, *Malthusian Moment.*

168 **Maynard remembers the "rush of dread" she felt** Joyce Maynard quoted in Hartmann, *America Syndrome.*

168 **Ehrlich became a celebrity with the stature** *The Tonight Show Starring Johnny Carson*, June 7, 1977; Ehrlich, interview by WOI-TV.

169 **Prestigious institutions showered Ehrlich with honors** Horowitz, *Anxieties of Affluence*; Hartmann, *America Syndrome.*

169 **Top Hollywood directors and actors signed on** Normandin and Valles, "How a Network of Conservationists."

170 **"teeming urban areas" where "a dangerous crisis . . ."** Robertson, *Malthusian Moment*, 181.

170 **critiques did not slow the population control movement's momentum** "The Population Bomb?" *New York Times*, May 31, 2015.

170 **Ehrlich announced the founding of a new organization** Wade Green, "The Militant Malthusians," *Saturday Review*, March 11, 1972; Robertson, *Malthusian Moment.*

172 **which shipped 1 million IUDs to India** Matthew Connelly, "Population Control in India: Prologue to the Emergency Period," *Population and Development Review* 32, no. 4 (2006): 629–67.

172 **population control movement scored a major victory** Gutiérrez, *Fertile Matters*; Charles Panati and Mary Lord, "Population Implosion," *Newsweek*, December 6, 1976; Henry Kamm, "India State Is Leader in Forced Sterilization," *New York Times*, August 13, 1976.

172 **Tanton, an unassuming man** Robertson, *Malthusian Moment*; "Dr. John Tanton—Founder of the Modern Immigration Network," John Tanton.org; Normandin and Valles, "How a Network of Conservationists"; Rohe, *Mary Lou and John Tanton.*

173 **parallels between the beehive and human population** Robert W. Currie, "The Biology and Behaviour of Drones," *Bee World* 68, no. 3 (1987): 129–43,

https://doi.org/10.1080/0005772X.1987.11098922; Elizabeth Anne Brown, "How Humans Are Messing Up Bee Sex," *National Geographic*, September 11, 2018; Social Contract, "Tribute to Tanton."

174 **Hardin had used a similarly misleading metaphor** Garrett Hardin, "Commentary: Living on a Lifeboat," *BioScience* 24, no. 10 (1974): 561–68; Constance Holden, "'Tragedy of the Commons' Author Dies," *Science*, September 26, 2003; Ehrlich and Holdren, "Impact of Population Growth."

174 **Tanton prided himself on his emotional detachment** Social Contract, "Tribute to Tanton"; Rohe, *Mary Lou and John Tanton*.

174 **He characterized them as separate species** Robertson, *Malthusian Moment*; Normandin and Valles, "How a Network of Conservationists"; Miriam King and Steven Ruggles, "American Immigration, Fertility, and Race Suicide at the Turn of the Century," *Journal of Interdisciplinary History* (Winter 1990); Southern Poverty Law Center, "John Tanton," https://www.splcenter.org /fighting-hate/extremist-files/individual/john-tanton.

175 **"How could *homo contraceptivus* compete . . ."** Social Contract, "Tribute to Tanton."

175 **Environmentalists in the population control movement** Michael Egan, *Barry Commoner and the Science of Survival: The Remaking of American Environmentalism* (Cambridge, MA: MIT Press, 2014); Ronald Bailey, "Real Environmental Racism," Reason.com, March 5, 2003.

176 **Tanton quickly ascended** Robertson, *Malthusian Moment*.

176 **"role of international migration in perpetuating . . ."** Tanton, "International Migration."

176 **exposé of India's population control program** Lewis M. Simons, "Compulsory Sterilization Provokes Fear, Contempt," *Washington Post*, July 4, 1977; Henry Kamm, "India State Is Leader in Forced Sterilization," *New York Times*, August 13, 1976; C. Brian Smith, "In 1976, More Than 6 Million Men in India Were Coerced into Sterilization," *Mel*, undated, https://melmagazine.com/en -us/story/in-1976-more-than-6-million-men-in-india-were-coerced-into -sterilization.

177 **a dozen states considered passing laws** Dennis Hodgson, "Orthodoxy and Revisionism in American Demography," *Population and Development Review* 14, no. 4 (1988): 541–69.

177 **Outraged feminists attacked Ehrlich as the mastermind** Robertson, *Malthusian Moment*.

178 **"I was trying to get something done"** Mark Malkoff, *The Carson Podcast with Guest Dr. Paul Ehrlich*, April 12, 2018.

178 **By the time the population control movement crashed** Gutiérrez, *Fertile Matters*; Mikko Myrskylä, Hans-Peter Kohler, and Francesco C. Billari, "Advances in Development Reverse Fertility Declines," *Nature* 460, no. 7256 (2009): 741.

178 **Activists concerned about the state of the environment** Anne Hendrixson, "Population Control in the Troubled Present: The '120 by 20' Target and Implant Access Program," *Development and Change* 50, no. 3 (2019): 786–804; Betsy Hartmann to author, July 26, 2019; Robertson, *Malthusian Moment*, 178.

179 **Allee, who had documented the positive effects of population density** Warder Clyde Allee, *The Social Life of Animals* (New York: W. W. Norton, 1938).

179 **what would later be hailed as "unambiguous" experiments** Franck Courchamp, Ludek Berec, and Joanna Gascoigne, *Allee Effects in Ecology and Conservation* (New York: Oxford University Press, 2008); Andrew T. Domondon, "A History of Altruism Focusing on Darwin, Allee and E. O. Wilson," *Endeavor*, June 2013.

180 **Bringing individuals together, in other words** Daniel Simberloff and Leah Gibbons, "Now You See Them, Now You Don't!—Population Crashes of Established Introduced Species," *Biological Invasions* 6, no. 2 (2004): 161–72; Desrochers and Hoffbauer, "Postwar Intellectual Roots."

181 **The innovative capacity of groups of people** Ehrlich and Holdren, "Impact of Population Growth"; Ramsden and Adams, "Escaping the Laboratory."

182 **Tanton spun off ZPG's immigration committee** Normandin and Valles, "How a Network of Conservationists"; DeParle, "Anti-Immigration Crusader"; "Anne H. Ehrlich," Wikipedia, https://en.wikipedia.org/wiki/Anne_H._Ehrlich; Rohe, *Mary Lou and John Tanton*.

183 **Tanton gently helped his supporters disregard** Tanton, "International Migration"; Social Contract, "Tribute to Tanton."

183 **It worked, for a while** Normandin and Valles, "How a Network of Conservationists"; DeParle, "Anti-Immigration Crusader."

183 **Brower and a faction of anti-immigration activists** Leon Kolankiewicz, "Homage to Iconic Conservationist David Brower Omits Population," Californians for Population Stabilization, March 25, 2014, https://www.capsweb.org/blog/homage-iconic-conservationist-david-brower-omits-population.

184 ***The Camp of the Saints*** Cécile Alduy, "What a 1973 French Novel Tells Us About Marine Le Pen, Steve Bannon, and the Rise of the Populist Right," *Politico*, April 23, 2017; Normandin and Valles, "How a Network of Conservationists"; K. C. McAlpin, " 'The Camp of the Saints' Revisited—Modern Critics

Have Justified the Message of a 1973 Novel on Mass Immigration," *Social Contract Journal*, Summer 2017.

184 **Tanton's organizations reconstructed the Fortress America** DeParle, "Anti-Immigration Crusader"; Normandin and Valles, "How a Network of Conservationists."

185 **a new social panic about migrants erupted** Allegra Kirkland, "Meet the Anti-Immigrant Crusader Trump Admin Tapped to Assist Immigrants," *Talking Points Memo*, May 1, 2017; Niraj Warikoo, "University of Michigan Blocks Release of Hot-Button Records of Anti-Immigrant Leader," *Detroit Free Press*, October 17, 2017; Eric Hananoki, "An Anti-Immigrant Hate Group Lobbying Director Is Now a Senior Adviser at US Citizenship and Immigration Services," *Media Matters for America*, March 7, 2018.

185 **In 2018 thirty-two of the thirty-four representatives** "NumbersUSA endorses Sen. Jeff Sessions for Attorney General," NumbersUSA, January 3, 2017, https://www.numbersusa.com/news/numbersusa-endorses-sen-jeff-sessions-attorney-general; Gaby Orr and Andrew Restuccia, "How Stephen Miller Made Immigration Personal," *Politico*, April 22, 2019. See also Leah Nelson, "NumbersUSA Denies Bigotry But Promotes Holocaust Denier," Southern Poverty Law Center, May 25, 2011, https://www.splcenter.org/hatewatch/2011/05/25/numbersusa-denies-bigotry-promotes-holocaust-denier.

186 **"You have to have the right genes . . ."** D'Antonio, "Trump's Move."

186 **"I don't believe in this doctrine of racial equality"** Liam Stack, "Holocaust Denier Is Likely GOP Nominee in Illinois," *New York Times*, February 8, 2018.

186 **They implied that mixing biologically distinct peoples** Gavin Evans, "The Unwelcome Revival of 'Race Science,'" *Guardian*, March 2, 2018; Nicole Hemmer, "'Scientific Racism' Is on the Rise on the Right. But It's Been Lurking There for Years," *Vox*, March 28, 2017; D'Antonio, "Trump's Move."

187 **allegedly claimed that "if we can get rid of enough people . . ."** Alexander C. Kaufman, "El Paso Terrorism Suspect's Alleged Manifesto Highlights Eco-Fascism's Revival," *HuffPost*, August 4, 2019.

187 **"Let them call you racists . . ."** Adam Nossiter, "'Let Them Call You Racists': Bannon's Pep Talk to National Front," *New York Times*, March 10, 2018.

## 7: HOMO MIGRATIO

191 **According to his "Aryan Polynesian" theory** Doug Herman, "How the Voyage of the Kon-Tiki Misled the World About Navigating the Pacific," *Smithsonian*, September 4, 2014; Finney, "Myth, Experiment."

191 **Peoples in Polynesia had an alternative theory** Lewis, *We, the Navigators.*

192 **Other canoe-traveling migrations had followed** S. H. Riesenberg, foreword to Lewis, *We, the Navigators.*

192 **did not accept the Great Fleet theory** Holton, "Heyerdahl's Kon Tiki Theory"; Ben Finney, "Founding the Polynesian Voyaging Society," *From Sea to Space* (Palmerston North, NZ: Massey University, 1992).

193 **Heyerdahl's Kon-Tiki theory presupposed an improbable journey** "About Thor Heyerdahl," Kon-Tiki Museum, https://www.kon-tiki.no/thor-heyerdahl/; "Kon-Tiki (1947)," Kon-Tiki Museum, https://www.kon-tiki.no/expeditions /kon-tiki-expedition/; John Noble Wilford, "Thor Heyerdahl Dies at 87; His Voyage on Kon-Tiki Argued for Ancient Mariners," *New York Times*, April 19, 2002.

194 **into the brisk waters off the Peruvian coast** "Scientists Meet Storm," *New York Times*, July 8, 1947; Thor Heyerdahl, "Kon-tiki Men Feel Safe, 6 Weeks Out," *New York Times*, July 7, 1947; "Parrot Vanishes as Gale Whips Kon-Tiki Raft," *New York Times*, July 9, 1947.

194 **Heyerdahl wrote a book about the Kon-Tiki journey** Holton, "Heyerdahl's Kon Tiki Theory."

195 **The most salient objection to Heyerdahl's Kon-Tiki theory** Finney, "Myth, Experiment."

196 **He claimed that there'd been no prehistoric migrations at all** Montagu, "What Is Remarkable."

196 **Scientific belief in biologically distinct racial groups** Marcos Chor Maio and Ricardo Ventura Santos, "Antiracism and the Uses of Science in the Post-World War II: An Analysis of UNESCO's First Statements on Race (1950 and 1951)," *Vibrant: Virtual Brazilian Anthropology* 12, no. 2 (2015): 1–26; Provine, "Geneticists and the Biology"; Michelle Brattain, "Race, Racism, and Antiracism: UNESCO and the Politics of Presenting Science to the Postwar Public," *American Historical Review* 112, no. 5 (2007): 1386–413, www.jstor.org/stable /40007100.

197 **the still-powerful fantasy of a racial order** Montagu, "What Is Remarkable."

198 **"bold and imaginative" and of "major scientific importance"** Ernst Mayr, "*Origin of the Human Races* by Carleton Coon" (review), *Science* (October 19, 1962): 420–22.

198 **Because the idea of a racial order in nature** Dobzhansky, "Possibility that *Homo sapiens.*"

198 **Plus, Coon's theory conflicted** Montagu, "What Is Remarkable."

199 **"practiced racial segregation during their wanderings"** Dobzhansky, "Possibility that *Homo sapiens.*"

199 **Civil rights activists condemned Coon's theory** John P. Jackson, "'In Ways Unacademical': The Reception of Carleton S. Coon's *The Origin of Races*," *Journal of the History of Biology* 34 (2001): 247–85; Dobzhansky, "Possibility that *Homo sapiens*."

199 **Decades would pass before scientists recovered** Vincent M. Sarich and Allan C. Wilson, "Immunological Time Scale for Hominid Evolution," *Science* 158, no. 3805 (1967): 1200–1203.

201 **for a study of their mitochondrial DNA** John Tierney and Lynda Wright, "The Search for Adam and Eve," *Newsweek*, January 11, 1988.

202 **Much more variation existed between individuals** Richard Lewontin, "The Apportionment of Human Diversity," *Evolutionary Biology* 6 (1972): 381–98.

202 **commentators viewed the notion of a mass migration out of Africa with suspicion** Alan G. Thorne and Milford H. Wolpoff, "The Multiregional Evolution of Humans," *Scientific American*, April 1992; Marek Kohn, "All About Eve and Evolution," *Independent*, May 3, 1993.

202 **incorporated the new DNA evidence** Jun Z. Li et al., "Worldwide Human Relationships Inferred from Genome-Wide Patterns of Variation," *Science* 319, no. 5866 (2008): 1100–1104; Brenna M. Henn, L. Luca Cavalli-Sforza, and Marcus W. Feldman, "The Great Human Expansion," *Proceedings of the National Academy of Sciences* 109, no. 44 (2012): 17758–64.

203 **He pieced together the story of their ancestors' movements** Roberts, "How to Sample."

203 **Local communities targeted by his team** Sribala Subramanian, "The Story in Our Genes," *Time*, January 16, 1995; Amade M'charek, *The Human Genome Diversity Project: An Ethnography of Scientific Practice* (New York: Cambridge University Press, 2005).

205 **"I am very troubled . . ."** Roberts, "How to Sample"; Marks, *Human Biodiversity*, 124.

205 **a century and a half after Darwin had first proposed** Darwin, *Descent of Man*.

206 **Ever since the days of Weismannism** Marks, *Human Biodiversity*, 174; Roberts, *Fatal Invention*.

207 **about the same number as the lowly worm** Steven Rose, "How to Get Another Thorax," *London Review of Books*, September 8, 2016.

207 **"No one could have imagined"** Jyoti Madhusoodanan, "Human Gene Set Shrinks Again," *Scientist*, July 8, 2014.

207 **Studies of the genetics of our fellow primates** Roberts, *Fatal Invention*.

208 **Chimpanzees, primatologists had found** Wolfgang Enard and Svante Pääbo, "Comparative Primate Genomics," *Annual Review of Genomics and Human Genetics* 5 (2004): 351–78.

208 **Still, confronted with the new genetic evidence exposing the myth** Jonathan Marks, "Ten Facts about Human Variation," in M. Muehlenbein, ed., *Human Evolutionary Biology* (Cambridge: Cambridge University Press, 2010); Nicholas Wade, "Gene Study Identifies 5 Main Human Populations, Linking Them to Geography," *New York Times*, December 20, 2002.

209 **"genetic data show that races clearly do exist"** Armand Marie Leroi, "A Family Tree in Every Gene," *New York Times*, March 14, 2005; Reich, *Who We Are*, xii; David Reich, "How Genetics Is Changing Our Understanding of 'Race,'" *New York Times*, March 23, 2018.

209 **Myths about Linnaean-style biological difference** Kelly M. Hoffman et al., "Racial Bias in Pain Assessment and Treatment Recommendations, and False Beliefs About Biological Differences Between Blacks and Whites," *Proceedings of the National Academy of Sciences* 113, no. 16 (2016): 4296–301; Alexandria Wilkins, Victoria Efetevbia, and Esther Gross, "Reducing Implicit Bias, Raising Quality of Care May Reduce High Maternal Mortality Rates for Black Women," *Child Trends*, April 25, 2019.

210 **maps depicting human genetic variation** David López Herráez et al., "Genetic Variation and Recent Positive Selection in Worldwide Human Populations: Evidence from Nearly 1 Million SNPs," *PLOS One* 4, no. 11 (2009): e7888.

211 **"dogs and wolves are nearly identical at the genetic level . . ."** Roberts, *Fatal Invention*, 51.

211 **maps had anti-immigrant and white supremacist commentators crowing** Steve Sailer, "Cavalli-Sforza's Ink Cloud," Vdare.com, May 24, 2000; Samuel Francis, "The Truth About a Forbidden Subject," *San Diego Union-Tribune*, June 8, 2000.

211 **Policies that failed to recognize racial biology** Harmon, "Why White Supremacists"; Will Sommer, "GOP Congressmen Meet with Accused Holocaust Denier Chuck Johnson," *Daily Beast*, January 16, 2019.

212 **The petrous bone is named after** Morten Rasmussen et al., "The Genome of a Late Pleistocene Human from a Clovis Burial Site in Western Montana," *Nature* 506, no. 7487 (2014): 225; Ron Pinhasi et al., "Optimal Ancient DNA Yields from the Inner Ear Part of the Human Petrous Bone," *PLOS One* 10, no. 6 (2015): e0129102.

213 **the "mother lode" of ancient DNA** Reich, *Who We Are*.

213 **Ancient peoples, after their arrival** Joseph K. Pickrell and David Reich, "Toward a New History and Geography of Human Genes Informed by Ancient DNA," *Trends in Genetics* 30, no. 9 (2014): 377–89.

214 **According to new DNA analyses** Jane Qui, "The Surprisingly Early Settlement of the Tibetan Plateau," *Scientific American*, March 1, 2017.

214 **No freak accident deposited unsuspecting people** Reich, *Who We Are*, 201–3.

215 **We've been migrants all along** Henry Nicholls, "Ancient Swedish Farmer Came from the Mediterranean," *Nature*, April 26, 2012; Reich, *Who We Are*, xiv–xxii, 96.

215 **Botanists call the process "inosculation"** Peter C. Simms, "The Only Love Honored by the Gods—Inosculation," Garden of Gods and Monsters, September 12, 2014, https://gardenofgodsandmonsters.wordpress.com/2014/09/12/the-only-love-honored-by-the-gods-inosculation/.

216 **Linguistic, archaeological, and ancient DNA evidence** Ann Gibbons, "'Game-changing' Study Suggests First Polynesians Voyaged All the Way from East Asia," *Science*, October 3, 2016.

216 **Experts now widely recognize their migration** Finney, "Myth, Experiment"; Álvaro Montenegro, Richard T. Callaghan, and Scott M. Fitzpatrick, "Using Seafaring Simulations and Shortest-Hop Trajectories to Model the Prehistoric Colonization of Remote Oceania," *Proceedings of the National Academy of Sciences* 113, no. 45 (2016): 12685–90.

217 **During one attempted crossing in 2017** "Two Women Sailing from Hawaii to Tahiti Are Rescued After Five Months Lost in the Pacific," *Los Angeles Times*, October 27, 2017.

218 **completed nine voyages using traditional wayfinding** Lewis, *We, the Navigators*.

218 **The potato had made it across the Pacific on its own** Carl Zimmer, "All by Itself, the Humble Sweet Potato Colonized the World," *New York Times*, April 12, 2018.

## 8: THE WILD ALIEN

221 **"better served by something native"** Cape May Fall Festival, Cape May, NJ, October 21, 2017; Kristin Saltonstall, "Cryptic Invasion by a Non-Native Genotype of the Common Reed, *Phragmites australis*, into North America," *Proceedings of the National Academy of Sciences* 99, no. 4 (2002): 2445–49.

223 **There was no "reasonable geological evidence"** Queiroz, *Monkey's Voyage*, 26, 42, 112.

224 **the dilemma of how species had spread** Thompson, *Where Do Camels Belong?* 28; Queiroz, *Monkey's Voyage*, 41.

225 **Biogeographers started finding clues** Florian Maderspacher, "Evolution: Flight of the Ratites," *Current Biology* 27, no. 3 (2017): R110–R113; Thompson, *Where Do Camels Belong?* 12.

225 **Vicariance restored a "biological version of inertia"** Queiroz, *Monkey's Voyage*, 65, 86, 234.

226 **The few biogeographers who believed** Paul P. A. Mazza, "Pushing Your Luck," review of *Monkey's Voyage*, *BioScience*, May 2014; Robert H. Cowie and Brenden S. Holland, "Dispersal Is Fundamental to Biogeography and the Evolution of Biodiversity on Oceanic Islands," *Journal of Biogeography* 33 (2006): 193–98.

226 **No jackrabbit ever had** John H. Prescott, "Rafting of Jack Rabbit on Kelp," *Journal of Mammalogy* 40, no. 3 (1959): 443–44.

227 **national parks as oases** Alfred Runte, *National Parks: The American Experience* (Lincoln, NE: University of Nebraska Press, 1997), 179.

227 **The government extended those protections** "Executive Order 13112–1. Definitions," US Department of Agriculture National Invasive Species Information Center, https://www.invasivespeciesinfo.gov/executive-order-13112 -section-1-definitions.

227 **Three new subdisciplines emerged** Mark Davis, "Defining Nature. Competing Perspectives: Between Nativism and Ecological Novelty," *Mètode Science Studies Journal—Annual Review* 9 (2019).

227 **The pace of the onslaught was "unprecedented"** Mooney and Cleland, "Evolutionary Impact"; Chew, "Ending with Elton."

228 **the case against wildlife on the move** Warren, "Perspectives on 'Alien.'"

228 **native inhabitants would be ravaged by an onslaught** "Invasive Species," Hawaii Invasive Species Council, http://dlnr.hawaii.gov/hisc/info/.

228 **The interbreeding was "massive"** Mooney and Cleland, "Evolutionary Impact."

229 **pushing checkerspots to the edge of extinction** Rudi Mattoni et al., "The Endangered Quino Checkerspot Butterfly, *Euphydryas editha quino* (Lepidoptera: Nymphalidae)," *Journal of Research on the Lepidoptera* 34 (1997): 99–118, 1995.

229 **second-largest threat to biodiversity** Mooney and Cleland, "Evolutionary Impact"; Thompson, *Where Do Camels Belong?* 46, 108, 195–96; Warren, "Perspectives on 'Alien.'"

230 **suggested moving some threatened checkerspot colonies** "G2: Animal Rescue: How Can We Save Some of Our Most Charismatic Animals from Extinction Due to Climate Change? One US Biologist, Camille Parmesan, Has a Radical Suggestion: Just Pick Them Up and Move Them," *Guardian*, February 12, 2010.

230 **The intruders had to be eradicated** Stanley A. Temple, "The Nasty Necessity: Eradicating Exotics," *Conservation Biology* 4, no. 2 (1990): 113–15.

230 **ancient biogeographical borders had been transgressed** Rebecca Ostertag, interview by author, February 20, 2018; Rebecca Ostertag et al., "Ecosystem and Restoration Consequences of Invasive Woody Species Removal in

Hawaiian Lowland Wet Forest," *Ecosystems* 12, no. 3 (2009): 503–15; "Two New Species of Fungi that Kill 'Ōō Trees Get Hawaiian Names," *University of Hawai'i News*, April 16, 2018, https://www.hawaii.edu/news/2018/04/16/ohia -killing-fungi-get-hawaiian-names/.

232 **relegated to the "margins of ecological research"** Roland Kays et al., "Terrestrial Animal Tracking as an Eye on Life and Planet," *Science* 348, no. 6240 (2015): aaa2478.

232 **many dramatic and long-distance movements** "The Worldwide Migration Pattern Of White Storks: Differences and Consequences," Max Planck Institute for Ornithology, https://www.orn.mpg.de/2137470/the-worldwide-migration -pattern-of-white-storks-differences-and-consequences; "Saw-whet owl migration," Ned Smith Center for Nature and Art, http://www.nedsmithcenter .org/saw-whet-owl-migration/

232 **The now-famous migration of monarch butterflies** Bernd Heinrich, *The Homing Instinct: Meaning and Mystery in Animal Migration* (Boston: Mariner Books, 2015), 45.

233 **checkerspots had a "remarkable lack of wanderlust"** Paul R. Ehrlich, "Intrinsic Barriers to Dispersal in Checkerspot Butterfly," *Science*, July 14, 1961.

234 **"It cost 3,500 dollars"** Martin Wikelski, interview by author, September 7, 2017.

234 **"We'd have to physically go near the elephant"** Cheshire and Uberti, *Where the Animals Go*, 36.

234 **The U.S. military had a much better system** Mark Sullivan, "A Brief History of GPS," TechHive, August 9, 2012, https://www.pcworld.com/article /2000276/a-brief-history-of-gps.html.

236 **The long, curving trunk of the highland tamarind tree** Jean-Jacques Segalen, "Acacia heterophylla," Dave's Garden, February 15, 2016, https:// davesgarden.com/guides/articles/acacia-heterophylla.

237 **The likeness between the two species puzzled botanists** Johannes J. Le Roux et al., "Relatedness Defies Biogeography: The Tale of Two Island Endemics (*Acacia heterophylla* and *A. koa*)," *New Phytologist* 204, no. 1 (2014): 230–42; "Botanists Solve Tree Mystery," IOL.co.za, June 27, 2014, https://www .iol.co.za/dailynews/opinion/botanists-solve-tree-mystery-1710429.

237 **Botanists settled on two equally unsatisfying explanations** Marris, "Tree Hitched a Ride."

238 **Vicariance theory attributed the separation of monkey species** Queiroz, *Monkey's Voyage*, 166–67, 212–13, 293.

239 **"Giant flukes happen"** Marris, "Tree Hitched a ride."

239 **Defense Department stopped adding a jitter to its GPS** "Frequently Asked Questions About Selective Availability: Updated October 2001," GPS.gov, https://www.gps.gov/systems/gps/modernization/sa/faq/.

239 **allowed people to track the once-undetectable movements** Cheshire and Uberti, *Where the Animals Go.*

240 **"we find totally amazing new information . . ."** Wikelski interview.

241 **Wild species regularly roam beyond the borders** Cheshire and Uberti, *Where the Animals Go*; Queiroz, *Monkey's Voyage*, 148; Wikelski interview; Roland Kays et al., "Terrestrial Animal Tracking as an Eye on Life and Planet," *Science* 348, no. 6240 (2015): aaa2478; Kessler, "Most Extreme Migration?"

242 **Even the spiders** Dingle, *Migration*, 62.

243 **"Humans trying to achieve this . . ."** Iain Couzin, interview by author, August 25, 2017.

243 **They called the new field "movement ecology"** Cheshire and Uberti, *Where the Animals Go.*

244 **a new phase in human understanding** "Ears for Icarus: Russian Rocket Delivers Antenna for Animal Tracking System to the International Space Station," Max-Planck-Gesellschaft, February 13, 2018, https://www.mpg.de /11939385/ears-for-icarus.

244 **"We see the whole network of animals . . ."** Wikelski interview.

245 **ecologists have started to reevaluate their theories** Warren, "Perspectives on 'Alien.'"

245 **zero extinctions among the locals** Mooney and Cleland, "Evolutionary Impact."

245 **arrival of newcomers increases biodiversity** Mark Vellend et al., "Global Meta-Analysis Reveals No Net Change in Local-Scale Plant Biodiversity Over Time," *Proceedings of the National Academy of Sciences* 110, no. 48 (2013): 19456–59; Thompson, *Where Do Camels Belong?* 108.

246 **compared the impact** Mooney and Cleland, "Evolutionary Impact"; Jessica Gurevitch and Dianna K. Padilla, "Are Invasive Species a Major Cause of Extinctions?" *Trends in Ecology and Evolution* 19, no. 9 (2004): 470–74; Thompson, *Where Do Camels Belong?* 78, 119.

246 **neither reduced diversity nor displaced native species** Claude Lavoie, "Should We Care About Purple Loosestrife? The History of an Invasive Plant in North America," *Biological Invasions* 12, no. 7 (2010): 1967–99.

247 **Natives do, too** Thompson, *Where Do Camels Belong?* 46, 195–96.

247 **"is completely unrealistic"** Ostertag interview.

248 **Ostertag and Cordell devised a new experiment** Ostertag interview.

250 **folly of splitting wild creatures into natives and aliens** Thompson, *Where Do Camels Belong?* 2.

250 **"long and dynamic process of almost continuous reorganization"** Vladimir Torres et al., "Astronomical Tuning of Long Pollen Records Reveals the Dynamic History of Montane Biomes and Lake Levels in the Tropical High Andes During the Quaternary," *Quaternary Science Reviews* 63 (2013): 59–72.

9: THE MIGRANT FORMULA

253 **The tracks of the animals moving across the land** Jeff Parsons, interview by author, October 25, 2017; Scott A. Sherrill-Mix, Michael C. James, and Ransom A. Myers, "Migration Cues and Timing in Leatherback Sea Turtles," *Behavioral Ecology* 19, no. 2 (2007): 231–36; R. T. Holmes et al., "Black-throated Blue Warbler (*Setophaga caerulescens*)," in P. G. Rodewald, ed., *The Birds of North America* (Ithaca, NY: Cornell Lab of Ornithology, 2017), https://doi.org/10.2173/bna.btbwar.03.

254 **What drives creatures to move?** Dingle, *Migration*, 252, 420.

255 **individuals that live in habitats exposed to change** Allison K. Shaw and Iain D. Couzin, "Migration or Residency? The Evolution of Movement Behavior and Information Usage in Seasonal Environments," *American Naturalist* 181, no. 1 (2012): 114–24.

256 **migration is likely to emerge** Dingle, *Migration*, 22.

256 **And so as the northern hemisphere tilts** Dingle, *Migration*, 157; Christopher G. Guglielmo, "Obese Super Athletes: Fat-Fueled Migration in Birds and Bats," *Journal of Experimental Biology* 221, suppl. 1 (2018): jeb165753.

256 **a restlessness sets in** Dingle, *Migration*, 138–39.

257 **Wild animals' sensitivity to environmental perturbations** Elke Maier, "A Four-Legged Early-Warning System," ICARUS: Global Monitoring with Animals, https://www.icarus.mpg.de/11706/a-four-legged-early-warning-system.

259 **"probably responsible for much of the diversity . . ."** Martin Wikelski, interview by author, September 7, 2017; Richard A. Holland, et al., "The Secret Life of Oilbirds: New Insights into the Movement Ecology of a Unique Avian Frugivore," *PLOS One* 4, no. 12 (2009): e8264.

259 **Ecologists saw the dramatic effects in a population of wolves** Christine Mlot, "Are Isle Royale's Wolves Chasing Extinction?" *Science*, May 24, 2013.

260 **those that facilitate animal movements flourish** Joshua J. Tewksbury et al., "Corridors Affect Plants, Animals, and Their Interactions in Fragmented

Landscapes," *Proceedings of the National Academy of Sciences* 99, no. 20 (2002): 12923–26.

261 **when conditions are good, they can sharpen their genetic adaptations** Stu Weiss, interview by author, March 7, 2018.

261 **Just what might have triggered their movements** Marjo Saastamoinen et al., "Predictive Adaptive Responses: Condition-Dependent Impact of Adult Nutrition and Flight in the Tropical Butterfly *Bicyclus anynana*," *American Naturalist* 176, no. 6 (2010): 686–98; Dingle, *Migration*, 61.

262 **Somehow butterfly pioneers had emerged** Camille Parmesan, interview by author, January 7, 2018; GrrlScientist, "The Evolutionary Trap That Wiped Out Thousands of Butterflies," *Forbes*, May 9, 2018; J. S. Kennedy, "Migration, Behavioral and Ecological," in Mary Ann Rankin and Donald E. Wohlschlag, eds., *Contributions in Marine Science*, vol. 27 *Supplement* (1985); Paul R. Ehrlich et al., "Extinction, Reduction, Stability and Increase: The Responses of Checkerspot Butterfly (Euphydryas) Populations to the California Drought," *Oecologia* 46, no. 1 (1980): 101–5; Susan Harrison, "Long-Distance Dispersal and Colonization in the Bay Checkerspot Butterfly, *Euphydryas editha bayensis*," *Ecology* 70, no. 5 (1989): 1236–43, www.jstor.org/stable/1938181.

265 **motives and impact of human migration remain shadowy** Jablonski, *Living Color*, 42.

265–66 **We moved in sync with the animals** Timothy P. Foran, "Economic Activities: Fur Trade," Virtual Museum of New France, Canadian Museum of History, https://www.historymuseum.ca/virtual-museum-of-new-france/economic-activities/fur-trade/; Marc Larocque, "Whaling, Overpopulation of Azores Led to Portuguese Immigration to SouthCoast," *Herald News* (Fall River, MA), June 10, 2012.

266 **Most every migrant could accurately describe** Mavroudi and Nagel, *Global Migration*, 99; "Remittances," Migration Data Portal, International Organization on Migration Global Migration Data Analysis Centre, https://migrationdataportal.org/themes/remittances#key-trends.

266 **And it doesn't actually work** Douglas S. Massey et al., "Theories of International Migration: A Review and Appraisal," *Population and Development Review* 19, no. 3 (1993): 431–66.

267 **Other popular theories** See, e.g., Crawford and Campbell, *Causes and Consequences*; Mukherjee, *Gene*, 339.

267 **a last-ditch one, forced by catastrophe** McLeman, *Climate and Human Migration*; Etienne Piguet, "From 'Primitive Migration' to 'Climate Refugees': The Curious Fate of the Natural Environment in Migration Studies," *Annals of the*

*Association of American Geographers* 103 (2013): 148–62; Issie Lapowsky, "How Climate Change Became a National Security Problem," *Wired*, October 20, 2015; Peter B. DeMenocal, "Cultural Responses to Climate Change During the Late Holocene," *Science* 292, no. 5517 (2001): 667–73.

267 **during periods of opportunity, not crisis** Axel Timmermann and Tobias Friedrich, "Late Pleistocene Climate Drivers of Early Human Migration," *Nature* 538, no. 7623 (2016): 92.

270 **only New Zealand has considered the idea** Charlotte Edmond, "5 Places Relocating People Because of Climate Change," World Economic Forum, June 29, 2017; Charles Anderson, "New Zealand Considers Creating Climate Change Refugee Visas," *Guardian*, October 31, 2017.

270 **do not qualify** Karen Musalo, "Systematic Plan to Narrow Humanitarian Protection: A New Era of US Asylum Policy," 15th Annual Immigration Law and Policy Conference, Georgetown University Law Center, Washington, D.C., October 1, 2018.

271 **They'd be sent back** Lauren Carasik, "Trump's Safe Third Country Agreement with Guatemala Is a Lie," *Foreign Policy*, July 30, 2019.

271 **"rooted in a sedentarist notion"** Richard Black et al., "The Effect of Environmental Change on Human Migration," *Global Environmental Change*, December 2011.

271 **telling evidence suggests that migration is encoded** Jonathan K. Pritchard, "How We Are Evolving," *Scientific American*, December 7, 2012; Carl Zimmer, "Genes for Skin Color Rebut Dated Notions of Race, Researchers Say," *New York Times*, October 12, 2017.

272 **The frequency of genes** D. Peter Snustad and Michael J. Simmons, *Principles of Genetics*, 6th ed. (Hoboken, NJ: John Wiley & Sons, 2012); I. Lobo, "Environmental Influences on Gene Expression," *Nature Education* 1, no. 1 (2008): 39; Patrick Bateson et al., "Developmental Plasticity and Human Health," *Nature* 430, no. 6998 (2004): 419.

272 **etching on our hands** Michael Kücken and Alan C. Newell, "Fingerprint Formation," *Journal of Theoretical Biology* 235, no. 1 (2005): 71–83.

273 **Their bodies absorbed signals of famine** Carl Zimmer, "The Famine Ended 70 Years Ago, But Dutch Genes Still Bear Scars," *New York Times*, January 31, 2018; Peter Ekamper et al., "Independent and Additive Association of Prenatal Famine Exposure and Intermediary Life Conditions with Adult Mortality Between Age 18–63 Years," *Social Science and Medicine* 119 (2014): 232–39.

273 **environmental conditions shape the development of our bodies** David J. P. Barker, "The Origins of the Developmental Origins Theory," *Journal of Internal Medicine* 261, no. 5 (2007): 412–17.

274 **People who lived around the Ganges** J. B. Harris et al. "Susceptibility to *Vibrio cholerae* Infection in a Cohort of Household Contacts of Patients with Cholera in Bangladesh," *PLOS Neglected Tropical Diseases* 2 (2008): e221.

274 **When the weak sunlight of northern climes** A. W. C. Yuen and N. G. Jablonski, "Vitamin D: In the Evolution of Human Skin Colour," *Medical Hypotheses* 74, no. 1 (2010): 39–44.

274 **Those who moved into cold regions** William R. Leonard et al., "Climatic Influences on Basal Metabolic Rates Among Circumpolar Populations," *American Journal of Human Biology* 14, no. 5 (2002): 609–20; Caleb E. Finch and Craig B. Stanford, "Meat-adaptive Genes and the Evolution of Slower Aging in Humans," *Quarterly Review of Biology* 79, no. 1 (2004): 3–50; Kumar S. D. Kothapalli et al., "Positive Selection on a Regulatory Insertion–Deletion Polymorphism in FADS2 Influences Apparent Endogenous Synthesis of Arachidonic Acid," *Molecular Biology and Evolution* 33, no. 7 (2016): 1726–39; Harmon, "Why White Supremacists"; Pascale Gerbault et al., "Evolution of Lactase Persistence: An Example of Human Niche Construction," *Philosophical Transactions of the Royal Society B: Biological Sciences* 366, no. 1566 (2011): 863–77.

275 **People from Tibet to this day** Mark Aldenderfer, "Peopling the Tibetan Plateau: Migrants, Genes and Genetic Adaptations," in Crawford and Campbell, *Causes and Consequences.*

275 **Our bodies' adaptations** Aneri Pattani, "They Were Shorter and at Risk for Arthritis, But They Survived an Ice Age," *New York Times*, July 6, 2017; Jacob J. E. Koopman et al., "An Emerging Epidemic of Noncommunicable Diseases in Developing Populations Due to a Triple Evolutionary Mismatch," *American Journal of Tropical Medicine and Hygiene* (2016): 1189–92; Isabelle C. Withrock et al., "Genetic Diseases Conferring Resistance to Infectious Diseases," *Genes and Diseases* 2, no. 3 (2015): 247–54; G. Genovese et al., "Association of Trypanolytic *ApoL1* Variants with Kidney Disease in African Americans," *Science* 329 (2010): 841–45.

276 **one potential candidate has been found** Benjamin C. Campbell and Lindsay Barone, "Evolutionary Basis of Human Migration," in Crawford and Campbell, *Causes and Consequences.*

276 **our bodies are fluid** Jonathon C. K. Wells and Jay T. Stock, "The Biology of Human Migration: The Ape that Won't Commit?" in Crawford and Campbell, *Causes and Consequences.*

277 **human migrants change the ecosystems they enter** McLeman, *Climate and Human Migration*; Nagel, *Global Migration*, 95; David P. Lindstrom and Adriana López Ramírez, "Pioneers and Followers: Migrant Selectivity and the Development of US Migration Streams in Latin America," *Annals of the American Academy*

*of Political and Social Science* 630, no. 1 (2010): 53–77; Alexander Domnich et al., "The 'Healthy Immigrant' Effect: Does It Exist in Europe Today?," *Italian Journal of Public Health* 9, no. 3 (2012); Steven Kennedy et al., "The Healthy Immigrant Effect: Patterns and Evidence from Four Countries," *Journal of International Migration and Integration* 16, no. 2 (2015): 317–32.

278 **In one study of immigrants in the United States** See, e.g., National Academies of Sciences, Engineering, and Medicine, and Committee on Population, *The Integration of Immigrants Into American Society* (National Academies Press, 2016); Francine D. Blau et al., "The Transmission of Women's Fertility, Human Capital, and Work Orientation Across Immigrant Generations," *Journal of Population Economics* 26, no. 2 (2013): 405–35.

279 **will undoubtedly be disruptive to communities** Sohini Ramachandran and Noah A. Rosenberg, "A Test of the Influence of Continental Axes of Orientation on Patterns of Human Gene Flow," *American Journal of Physical Anthropology* 146, no. 4 (2011): 515–29.

279 **There is no straightforward equation** McLeman, *Climate and Human Migration.*

279 **One kind of environmental change** Richard Black et al., "The Effect of Environmental Change on Human Migration," *Global Environmental Change*, December 2011; Etienne Piguet, Antoine Pécoud, and Paul de Guchteneire, "Introduction: Migration and Climate Change," in Etienne Piguet et al., eds., *Migration and Climate Change* (New York: Cambridge University Press, 2011), 9; McLeman, *Climate and Human Migration*; Dina Ionesco, Daria Mokhnacheva, and François Gemenne, *The Atlas of Environmental Migration* (London: Routledge, 2016); Anastasia Moloney, "Two Million Risk Hunger After Drought in Central America," Reuters, September 7, 2018; Lauren Markham, "The Caravan Is a Climate Change Story," *Sierra*, November 9, 2018.

280 **The mass exodus out of Syria** Colin P. Kelley et al., "Climate Change in the Fertile Crescent and Implications of the Recent Syrian Drought," *Proceedings of the National Academy of Sciences* 112, no. 11 (2015): 3241–46.

280 **But the drought alone did not cause** Helene Bie Lilleor and Kathleen Van den Broeck, "Economic Drivers of Migration and Climate Change in LDCs," *Global Environmental Change* 21S (2011), s70–81.

281 **picking up and leaving isn't the sole option** "Water Is 'Catalyst' for Cooperation, Not Conflict, UN Chief Tells Security Council," *UN News*, June 6, 2017; Philipp Blom, *Nature's Mutiny: How the Little Ice Age of the Long Seventeenth Century Transformed the West and Shaped the Present* (New York: W. W. Norton,

2017); John Lanchester, "How the Little Ice Age Changed History," *New Yorker*, April 1, 2019; United Nations Convention to Combat Desertification, "The Great Green Wall Initiative," https://www.unccd.int/actions/great-green-wall -initiative.

10: THE WALL

284 **He'd received many such bodies by then** Christos Mavrakidis, interview by author, June 2016.

285 **The remains of migrants who die in the desert** Gocha, Spradley, and Strand, "Bodies in Limbo"; Manny Fernandez, "A Path to America, Marked by More and More Bodies," *New York Times*, May 4, 2017.

286 **Against the odds, someone found** Kate Spradley and Eddie Canales, interview by author, January 8, 2018; Mark Reagan and Lorenzo Zazueta-Castro, "Death of a Dream: Hundreds of Migrants Have Died Crossing Into Valley," *Monitor* (McAllen, Tex.), July 28, 2019.

288 **The first thing they'd do was try to identify** Gocha, Spradley, and Strand, "Bodies in Limbo."

288 **geographic barriers prevent wild species from migrating as effectively** Michael T. Burrows et al., "Geographical Limits to Species-Range Shifts Are Suggested by Climate Velocity," *Nature* 507, no. 7493 (2014): 492; John R. Platt, "Climate Change Claims Its First Mammal Extinction," *Scientific American*, March 21, 2019.

289 **Our massive footprint** Michael Miller, "New UC Map Shows Why People Flee," *UC News*, November 15, 2018; Stuart L. Pimm et al., "The Biodiversity of Species and Their Rates of Extinction, Distribution, and Protection," *Science* 344, no. 6187 (2014): 1246752.

289 **Species that have not lost their habitats** Ken Wells, "Wildlife Crossings Get a Whole New Look," *Wall Street Journal*, June 20, 2017; "World's Largest Wildlife Corridor to Be Built in California," Ecowatch, September 27, 2015; Gabe Bullard, "Animals Like Green Space in Cities—And That's a Problem," *National Geographic*, April 20, 2016; Eliza Barclay and Sarah Frostenson, "The Ecological Disaster That Is Trump's Border Wall: A Visual Guide," *Vox*, February 5, 2019.

290 **the more constrained animal movements became** Marlee A. Tucker et al., "Moving in the Anthropocene: Global Reductions in Terrestrial Mammalian Movements," *Science* 359, no. 6374 (2018): 466–69.

290 **Animals, winds, currents, and waves could freely travel** Elisabeth Vallet, "Border Walls and the Illusion of Deterrence," in Jones, *Open Borders*; see also

Samuel Granados et al., "Raising Barriers: A New Age of Walls: Episode 1," *Washington Post*, October 12, 2016; David Frye, *Walls: A History of Civilization in Blood and Brick* (New York: Scribner, 2018), 238.

291 **Walls don't necessarily function** Noah Greenwald et al., "A Wall in the Wild: The Disastrous Impacts of Trump's Border Wall on Wildlife," Center for Biological Diversity, May 2017; Jamie W. McCallum, J. Marcus Rowcliffe, and Innes C. Cuthill, "Conservation on International Boundaries: The Impact of Security Barriers on Selected Terrestrial Mammals in Four Protected Areas in Arizona, USA," *Plos one* 9, no. 4 (2014): e93679.

291 **"is very much like squeezing a balloon . . ."** McAllister and Prentice, "African Migrants Turn to Deadly Ocean Route."

292 **Migration continues, but in a deadlier form** See Reece Jones, *Violent Borders: Refugees and the Right to Move* (London: Verso, 2016).

292 **one migrant died for every fifty-one who arrived** UNHCR, "Desperate Journeys: Refugees and Migrants Arriving in Europe and at Europe's Borders," January-December 2018.

292 **But the flow of people had simply shifted** Joe Penney, "Why More Migrants Are Dying in the Sahara," *New York Times*, August 22, 2017; McAllister and Prentice, "African Migrants Turn to Deadly Ocean Route."

293 **over 33,000 people died trying to migrate** Alan Cowell, "German Newspaper Catalogs 33,293 Who Died Trying to Enter Europe," *New York Times*, November 13, 2017.

293 **22,000 may have died trying to cross** The official number of deaths counted by U.S. Border Patrol between 1998 and 2018 is 7,505. See U.S. Border Patrol, "Southwest Border Sectors: Southwest Border Deaths by Fiscal Year," at https://www.cbp.gov/sites/default/files/assets/documents/2019-Mar/bp -southwest-border-sector-deaths-fy1998-fy2018.pdf. Experts agree this is an underestimate. A 2018 investigation by *USA Today* reporters estimates that the true number of deaths is between 25 and 300 percent higher. Rob O'Dell, Daniel González, and Jill Castellano, " 'Mass Disaster' Grows at the U.S.-Mexico Border, But Washington Doesn't Seem to Care," in "The Wall: Unknown Stories, Unintended Consequences," *USA Today* Network special report, 2018, https://www.usatoday.com/border-wall/.

293 **The true figures are probably much higher** Manny Fernandez, "A Path to America, Marked by More and More Bodies," *New York Times*, May 4, 2017.

293 **the governments of Europe changed their minds** "Schengen: Controversial EU Free Movement Deal Explained," BBC, April 24, 2016; Piro Rexhepi, "Europe Wrote the Book on Demonising Refugees, Long Before Trump Read It," *Guardian*, February 21, 2017.

294 **Others hung themselves** Lizzie Dearden, "Syrian Asylum Seeker 'Hangs Himself' in Greece Amid Warnings Over Suicide Attempts by Trapped Refugees," *Independent*, March 28, 2017.

294 **Volunteer doctors watched as suicide rates** Doctors Without Borders members, interview by author, June 12, 2016.

295 **across mountains and seas in hopes of a different future** Ghulam Haqyar, interview by author, June 12, 2016.

296 **ruled in 2011 that they amounted to torture** Court of Justice of the European Union, "According to Advocate General Trstenjak, Asylum Seekers May Not Be Transferred to Other Member States If They Could There Face a Serious Breach of the Fundamental Rights Which They Are Guaranteed Under the Charter of Fundamental Rights," Press Release, September 22, 2011

296 **Their crude logic** "The Truth About Migration," *New Scientist*, April 6, 2016.

296 **exclude people without official documents from services** Sarah Spencer and Vanessa Hughes, "Outside and In: Legal Entitlements to Health Care and Education for Migrants with Irregular Status in Europe," COMPAS: Centre on Migration, Policy & Society, University of Oxford, July 2015; Michele LeVoy and Alyna C. Smith, "PICUM: A Platform for Advancing Undocumented Migrants' Rights, Including Equal Access to Health Services," *Public Health Aspects of Migration in Europe*, WHO Newsletters, no. 8, March 2016; Marianne Mollmann, "A New Low: Stealing Family Heirlooms in Exchange for Protection," *Physicians for Human Rights*, December 16, 2015.

296 **steadily erodes** Cornelis J. Laban et al., "The Impact of a Long Asylum Procedure on Quality of Life, Disability and Physical Health in Iraqi Asylum Seekers in the Netherlands," *Social Psychiatry and Psychiatric Epidemiology* 43, no. 7 (2008): 507–15.

296, 98 **beyond deprivation to the purposeful infliction of trauma** Laura C. N. Wood, "Impact of Punitive Immigration Policies, Parent-Child Separation and Child Detention on the Mental Health and Development of Children," *BMJ Paediatrics Open* 2, no. 1 (2018); Dara Lind, "A New York Courtroom Gave Every Detained Immigrant a Lawyer. The Results Were Staggering," *Vox*, November 9, 2017; Michelle Brané and Margo Schlanger, "This Is What's Really Happening to Kids at the Border," *Washington Post*, May 30, 2018; Jacob Soboroff, "Emails Show Trump Admin Had 'No Way to Link' Separated Migrant Children to Parents," NBC News, May 1, 2019.

298 **Critics, noting the unsanitary and overcrowded conditions** David Shepardson, "Trump Says Family Separations Deter Illegal Immigration," Reuters,

October 13, 2018; Dara Lind, "Trump's DHS Is Using an Extremely Dubious Statistic to Justify Splitting Up Families at the Border," *Vox*, May 8, 2018.

299 **Thanks to a backlog** Brittany Shoot, "Federal Government Shutdown Could More Than Double Wait Time for Immigration Cases," *Fortune*, January 11, 2019; Brett Samuels, "Trump Rejects Calls for More Immigration Judges: 'We Have to Have a Real Border, Not Judges,'" *Hill*, June 19, 2018.

299 **policies to stymie migrants' right to claim asylum** American Immigration Council, "A Primer on Expedited Removal," July 22, 2019; Caitlin Dickerson et al., "Migrants at the Border: Here's Why There's No Clear End to Chaos," *New York Times*, November 26, 2018; Andrea Pitzer, "Trump's 'Migrant Protection Protocols' Hurt the People They're Supposed to Help," *Washington Post*, July 18, 2019; Migration Policy Institute, "Top 10 Migration Issues of 2019."

300 **Other policies targeted immigrants already settled** Jomana Karadsheh and Kareem Khadder, "'Pillar of the Community' Deported from US After 39 Years to a Land He Barely Knows," CNN, February 9, 2018; Jenna DeAngelis, "Simsbury Business Owners Who Are Facing Deportation to China, Speak Out," FOX 61, February 6, 2018; Michelle Goldberg, "First They Came for the Migrants," *New York Times*, June 11, 2018.

300 **Citizens would be subject to denaturalization** Aaron Rupar, "Why the Trump Administration Is Going After Low-Income Immigrants, Explained by an Expert," *Vox*, August 12, 2019; Seth Freed Wessler, "Is Denaturalization the Next Front in the Trump Administration's War on Immigration?" *New York Times Magazine*, December 19, 2018.

300 **Officials in the White House watched approvingly** Zach Hindin and Mario Ariza, "When Nativism Becomes Normal," *Atlantic*, May 23, 2016; Jonathan M. Katz, "What Happened When a Nation Erased Birth-Right Citizenship," *Atlantic*, November 12, 2018.

301 **Haitian neighborhoods emptied** Geralde Gabeau, interview by author, October 24, 2017; Cindy Carcamo, "In San Diego, Haitians Watch Community Countrymen Leave for Canada," *Los Angeles Times*, August 27, 2017.

301 **The migrants streamed north** Michelle Ouellette, interview by author, October 5, 2017; Catherine Tunney, "How the Safe Third Country Agreement Is Changing Both Sides of the Border," CBC News, April 1, 2017.

302 **"I'm screaming and no one is around for my rescue"** Eric Taillefer, interview by author, October 2, 2017; Jonathan Montpetit, "Mamadou's Nightmare: One Man's Brush with Crossing U.S.-Quebec Border," CBC News, March 13, 2017.

303 **with nothing except the clothes on their backs** Catherine Solyom, "Canadian Government, Others Discouraging Haitians in U.S. from Seeking Asylum Here," *Montreal Gazette*, August 14, 2017; Taillefer interview; Katherine Wilton, "Montreal Schools Preparing for Hundreds of Asylum Seekers," *Montreal Gazette*, August 22, 2017.

303 **Jean-Pierre and his family were among them** "Jean-Pierre," interview by author, October 26, 2017.

304 **As the myth of a sedentary past evaporates** Lim, Metzler, and Bar-Yam, "Global Pattern Formation"; David Norman Smith and Eric Hanley, "The Anger Games: Who Voted for Donald Trump in the 2016 Election, and Why?" *Critical Sociology* 44, no. 2 (2018): 195–212.

304 **One study suggests that xenophobic outbursts** Wesley Hiers, Thomas Soehl, and Andreas Wimmer, "National Trauma and the Fear of Foreigners: How Past Geopolitical Threat Heightens Anti-Immigration Sentiment Today," *Social Forces* 96, no. 1 (2017): 361–88; Lim, Metzler, and Bar-Yam, "Global Pattern Formation"; Margaret E. Peters, "Why Did Republicans Become So Opposed to Immigration? Hint: It's Not Because There's More Nativism," *Washington Post*, January 30, 2018.

304 **One telling study analyzed the counties and states** Thomas Edsall, "How Immigration Foiled Hillary," *New York Times*, October 5, 2017.

305 **the burden exacted by novel influx of newcomers** Adam Ozimek, Kenan Fikri, and John Lettieri, "From Managing Decline to Building the Future: Could a Heartland Visa Help Struggling Regions?" Economic Innovation Group, April 2019.

305 **Another possible explanation has to do with the optics** Jablonski, *Living Color*; Charles Stagnor, Rajiv Jhangiani, and Hammond Tarry, "Ingroup Favoritism and Prejudice," *Principles of Social Psychology*, 1st international ed., 2019, https://opentextbc.ca/socialpsychology/.

306 **the border between natives and migrants can be nebulous** Jie Zong, Jeanne Batalova, and Micayla Burrows, "Frequently Requested Statistics on Immigrants and Immigration in the United States," Migration Policy Institute, March 14, 2019; Michael B. Sauter, "Population Migration: These Are the Cities Americans Are Abandoning the Most," *USA Today*, September 18, 2018.

307 **this tendency may have evolved as an immune response** Alfred W. Crosby, "Virgin Soil Epidemics as a Factor in the Aboriginal Depopulation in America," *William and Mary Quarterly* 33, no. 2 (1976): 289–99; Sonia Shah, *The Fever: How Malaria Has Ruled Humankind for 500,000 Years* (New York: Farrar, Straus and Giroux, 2010), 65.

308 **ethnocentrist and xenophobic tendencies do seem to correlate** C. L. Fincher and R. Thornhill, "Parasite-stress Promotes In-Group Assortative Sociality: The Cases of Strong Family Ties and Heightened Religiosity," *Behavioral and Brain Sciences* 35, no. 2 (2012): 61–79; Sunasir Dutta and Hayagreeva Rao, "Infectious Diseases, Contamination Rumors and Ethnic Violence: Regimental Mutinies in the Bengal Native Army in 1857 India," *Organizational Behavior and Human Decision Processes* 129 (2015): 36–47.

308 **Fever is an ancient** Elspeth V. Best and Mark D. Schwartz, "Fever," *Evolution, Medicine and Public Health* 2014, no. 1 (2014): 92; Peter Nalin, "What Causes a Fever?" *Scientific American*, November 21, 2005.

308 **overestimated the proportion of immigrants in their countries** Directorate General for Communication, "Special Barometer 469: Integration of Immigrants in the European Union," European Commission, April 2018.

309 **They continue regardless of facts** Daniel J. Hopkins, John Sides, and Jack Citrin, "The Muted Consequences of Correct Information About Immigration," *Journal of Politics* 81, no. 1 (2019): 315–20; Eduardo Porter and Karl Russell, "Migrants Are on the Rise Around the World, and Myths About Them Are Shaping Attitudes," *New York Times*, June 20, 2018.

CODA: SAFE PASSAGE

311 **a path to a more secure future** "Refugee Resettlement Facts," UNHCR, February 2019, https://www.unhcr.org/en-us/resettlement-in-the-united-states.html.

312 **curtails life expectancy by three decades** Andrea K. Walker, "Baltimoreans Are as Healthy as Their Neighborhoods," *Baltimore Sun*, November 12, 2012.

313 **new conservation efforts are seeking to stitch** "Our Progress," Yellowstone to Yukon Conservation Initiative, n.d., https://y2y.net/vision/our-progress/our-progress; "Man-made Corridors," Conservation Corridor, n.d., https://conservationcorridor.org/corridors-in-conservation/man-made-corridors/; Tony Hiss, "Can the World Really Set Aside Half of the Planet for Wildlife?" *Smithsonian*, September 2014.

314 **The ability to move is no panacea** Ed Yong, "The Disturbing Walrus Scene in *Our Planet*," *Atlantic*, April 8, 2019; Michael P. Marchetti and Tag Engstrom, "The Conservation Paradox of Endangered and Invasive Species," *Conservation Biology* 30, no. 2 (2016): 434–37.

315 **United Nations' Global Compact** Right-wing populist leaders and governments pulled out of the compact, despite the fact that it is nonbinding and voluntary, including the United States, Australia, Brazil, and a number of eastern European countries. Frey Lindsay, "Opposition to the Global Compact

for Migration Is Just Sound and Fury," *Forbes*, November 13, 2018; "Portugal Approves Plan to Implement Global Compact on Migration," *Famagusta Gazette*, August 2, 2019; Lex Rieffel, "The Global Compact on Migration: Dead on Arrival?" Brookings Institution, December 12, 2018; Edith M. Lederer, "UN General Assembly Endorses Global Migration Accord," Associated Press, December 19, 2018.

316 **The militarized borders that bar human movement** Jones, *Open Borders*; John Washington, "What Would an 'Open Borders' World Actually Look Like?" *Nation*, April 24, 2019.

317 **The wall itself exudes death** Matthew Suarez, interview by author, March 6, 2018.

# INDEX

Note: page numbers in italics refer to figures.

# A NOTE ON THE AUTHOR

SONIA SHAH is a science journalist and the prizewinning author of *Pandemic: Tracking Contagions, from Cholera to Ebola and Beyond*, a finalist for the Los Angeles Times Book Prize, the New York Public Library Helen Bernstein Book Award for Excellence in Journalism, and the National Association of Science Writers' Science in Society Award. She has written for the *New York Times*, the *Wall Street Journal*, and many others. Her TED talk, "Three Reasons We Still Haven't Gotten Rid of Malaria," has been viewed by more than one million people around the world. She lives in Baltimore, Maryland.